Advances in Spatial Science

For further volumes:
http://www.springer.com/series/3302

Sven Erlander

Cost-Minimizing Choice Behavior in Transportation Planning

A Theoretical Framework for Logit Models

Springer

Professor Sven Erlander
Department of Mathematics
Linköping University
581 83 Linköping
Sweden
sverl@mai.liu.se

ISSN 1430-9602
ISBN 978-3-642-11910-1 e-ISBN 978-3-642-11911-8
DOI: 10.1007/978-3-642-11911-8
Springer Heidelberg Dordrecht London New York

Library of Congress Control Number: 2010927685

Cover design: eStudio Calamar, Spain

Printed on acid-free paper

Springer is part of Springer Science+Business Media (www.springer.com)

Front view

Staircase to second floor, Administration Building, Linköping University

Side view

Painting by Oscar Reutersvärd transformed into 3-dimensional sculpture by Björn Kruse and Tomas Ohlsson

Perspectives of Science

Preface

In the Administration building at Linköping University we have one of Oscar Reutersvärd's "Impossible Figures" in three dimensions. I call it "Perspectives of Science". When viewed from a specific point in space there is order and structure in the 3-dimensional figure. When viewed from other points there is disorder and no structure. If a specific scientific paradigm is used, there is order and structure; otherwise there is disorder and no structure.

My perspective in Transportation Science has focused on understanding the mathematical structure and the logic underlying the choice probability models in common use. My book with N. F. Stewart on the Gravity model (Erlander and Stewart 1990), was written in this perspective. The present book stems from the same desire to understand underlying assumptions and structure.

It investigates how far a new way of defining *Cost-Minimizing Behavior* can take us. It turns out that all commonly used choice probability distributions of logit type – log linear probability functions – follow from cost-minimizing behavior defined in the new way. In addition some new nested models appear.

The new way of defining cost-minimizing behavior is the following:
Cost-minimizing behavior obtains if the likelihood (probability) of any independent N-sample of observations is a decreasing function of the average cost of the N-sample. From this definition we have a basic formulation that can be falsified or tested by observations, thus satisfying the Popper criterion of the scientific status of a theory by its "falsifiability, or refutability or testability" (Popper 1963). The new way of defining cost-minimizing behavior is closely related to Tony Smith's *Cost-Efficiency Principle* (Smith 1978b, 1983, 1988).

In several books Amartya Sen, who received the Prize in Economic Sciences in Memory of Alfred Nobel in 2007, has investigated what should be meant by "freedom of choice" in deterministic cases. He writes. "The foundational importance of freedom of choice may well be the most far-reaching substantive problem neglected in standard economics" (Sen 1988). In this book I introduce a measure of freedom of choice in the probabilistic case related to the Shannon (1948) measure of how much "choice" is involved.

The standard approach in deriving choice probability functions is the Additive Random Utility Maximizing (ARUM) approach developed by Ben-Akiva (1974), Williams (1977), McFadden (1974, 1978a, 1981), and others. The standard approach uses composite cost – expected achieved perceived cost – as a welfare measure. This interpretation of the composite cost is not available in my approach. However, the interesting thing is that the same log sum formula comes out as a welfare measure combining cost and freedom of choice. This explains some of the strange behavior of the composite cost in limiting cases for the value of the parameter.

This book is intended for PhD students and researchers in Transportation Science, Transportation Planning, Regional Science, Regional Planning, Economics and Optimization. The reader should have a background in mathematics, optimization, probability, statistics, and economics, equivalent to a M Sc in Industrial Engineering and Management (I-linjen) at Linköping University. The style is mathematical with propositions and proofs. Technical details have been collected in an Appendix.

The book has grown out of my interest in and occupation with traffic planning models over many years. Many people have contributed to my knowledge and understanding of Transportation Science. I wish to thank D.E. Boyce, M. Florian, L.G. Mattsson, T.E. Smith and N.F. Stewart, for many interesting and inspiring discussions. In particular I wish to acknowledge the profound influence the ideas of Tony Smith have had on my thinking ever since we met and I read his pioneering paper on efficiency in 1977. My special thanks go also to Neil Stewart for the great fun we had in writing our book on the gravity model. I also wish to thank the faculty, in particular S. Danielsson and U. Hjorth, and former PhD students, P.Å. Andersson, K.O. Jörnsten, T. Larsson, J.T. Lundgren, M. Patriksson and S. Schele, at the Department of Mathematics, Linköping University. J.T. Lundgren joined me in writing early versions of a part of this book.

Linköping, Sweden
May 2010

Sven Erlander
Linköping University

Contents

Part II Equilibrium

Chapter 1
Logit Models for Spatial Interaction: Background

Abstract This book deals with logit type models which are commonly used in studying spatial interaction. The first approach we could call the Lagrangian paradigm. It is based on the formulation of a mathematical optimization problem the solutions of which have the desired properties, such as user equilibrium route choice and gravity model trip matrix. The second approach is the Additive Random Utility Maximizing ARUM paradigm. This book represents a third approach; a new definition of cost-minimizing behavior is used to derive and analyze logit type models for spatial interaction. This gives motivation for the models and admits verification of the structure of the models against observations. We start with an historical overview of different problems and the related development of ideas on how to formulate and solve the problems.

1.1 Introduction

This book deals with logit type probabilistic choice models, which are commonly used in studying spatial interaction, transportation planning in particular. There are two dominating approaches to the analysis of spatial interaction models of logit type.

The first approach we could call the Lagrangian paradigm. It is based on the formulation of a mathematical optimization problem the solutions of which have the desired properties, such as user equilibrium route choice. These models involve a gravity model trip matrix (Murchland 1966; Tomlin 1967; Wilson 1967; Florian et al. 1975; Florian et al. 1977; Evans 1976; Erlander 1977; Schéele 1977, 1980; Boyce 1980, 1984; Boyce et al. 1988). For a review see Boyce (2007).

The Additive Random Utility Maximizing ARUM paradigm is the second approach (Ben-Akiva 1974; Williams 1977; McFadden 1974, 1978a, 1981). For a review see McFadden (2001).

This book represents a third approach; a new definition of cost-minimizing behavior is used to derive and analyze logit type models for spatial interaction. This gives motivation for the models and admits verification of the structure of the models against observations. The third approach is essentially the efficiency principle

S. Erlander, *Cost-Minimizing Choice Behavior in Transportation Planning*, Advances in Spatial Science, DOI 10.1007/978-3-642-11911-8_1,

(Smith 1978a,b, 1983, 1988; Erlander 1985; Erlander and Smith 1990). We have however changed perspective. Here we use the central idea of the efficiency principle to *define* cost-minimizing behavior, and from this definition we derive all of our models.

We propose a welfare measure based on expected cost and expected freedom of choice inspired by the work by Sen (1988) in the deterministic case.

The main focus is on the assumptions underlying the models and on the properties of the models. A new way of deriving the models is given. The emphasis is on understanding the structure of the models and, in particular, what has to be assumed as given, and how these assumptions can be tested against observations. All logit type models obtained by the ARUM paradigm can be derived in the new way. Also many of the models obtained by the Lagrangian paradigm can be derived in the new way. The contribution of the new method of derivation is a better understanding of the structure of the models and a new foundation for the construction of new models. We analyze a number of commonplace models with the new approach and give a few examples of new models. In each chapter there is a discussion of the properties of the models with particular emphasis on motivation and verification.

We start with an historical overview of different problems and the related development of ideas on how to formulate and solve the problems. It will be seen that logit type models have developed in different fields and been applied in a wide range of applications.

Logit type models, or simply logit models, can be traced back to efforts to solve problems which in the transportation contexts may be labeled trip distribution, assignment of trips to a network, mode choice, and combinations of of these problems. Another line of development is in micro economic theory where logit models appear as econometric models.

There are three common elements in all analysis of spatial interaction. The first is the idea that decision makers/trip makers are rational. This means that the choices made are influenced by distances or costs and other factors in a more or less obvious way. Secondly, the number of decision makers/trip makers is large. It is often not suitable or possible to describe the decision making in a deterministic way for every individual decision maker/trip maker. Hence probabilistic models are being used. Thirdly, some kind of welfare measure is used to evaluate different solutions after a change, e.g., in cost structure.

By "cost" we can mean many things. Here we shall use cost as related to costs arising directly to the trip maker as a consequence of choosing a particular alternative. It may be money costs, travel times and also include other attributes that characterize the choice situation or the decision maker. This is often called "generalized cost".

Planning authorities use models to analyze and predict migration flows. What are the determinating factors influencing migration flows? What is the influence of housing, labour market, schools, health care, transportation, public transport, road system, parking, recreation areas?

Transportation itself is a particular field for analyze and prediction. How many trips will be generated between zones of a city? Distance and trip costs are obvious factors influencing trip making. How are the routes along the road net chosen? How are the trips divided between available modes of transportation such as private car, bus, train and underground? What will be the influence of the construction of new roads and increases of standard of public transport? Fees, tolls and taxes?

Localization problems combine factors of migration and transportation.

The first approaches to the questions above dealt with flows and numbers of individuals. The gravity model is one example. Later, efforts were made to understand and analyze the behavior of individual trip makers – decision makers. In this way "discrete choice" was introduced. This book represents one approach in this direction.

Logit type models have been widely used to study migration, transportation, land use, and telephone traffic.

The book is on theory. However, the empirical background and the real world problems should always be kept in mind.

1.2 Cost-Minimizing Behavior

> "This principle may in fact be viewed as a translation into macro terms of the basic behavioral hypothesis that, other things being equal, individuals tend to make shorter (less costly) trips rather than longer (more costly) trips. For if one adopts this hypothesis about individual trip behavior, then in observing macro trip activity within a system of interacting individuals, it is natural to hypothesize that from among all patterns of micro trip patterns consistent with this activity, those patterns involving lower total travel costs are more likely to have occurred than those involving higher costs." Smith (1978a)

An obvious way to describe the decision making of one individual is by assuming that her/his choice among possible alternatives can be described by a probability function. Also it seems safe to assume that, other things being equal, cheaper alternatives will be chosen more often than more expensive alternatives. Counterexamples to this rational behavior can be constructed, but it is no serious limitation to exclude them. Hence the probability of choosing an alternative will be a decreasing function of the cost of the alternative: the probability will decrease as the cost increases. In many applications of spatial interaction we have a population of decision makers. The population may be the residents of a city. We can then describe the choices made by one individual drawn at random from this population using a probability function of the type discussed above.

We can go one step further. Let us consider a group or, with the term used in statistics, a sample of size N of decision makers. How would cost-minimizing behavior be observed in this group or sample? What can be said about cost-minimizing behavior on the average in the group or sample?

For each member of the group or sample we can note the alternative chosen and the cost of this alternative. If the behavior of the members of the group/sample is cost minimizing we would expect to find lower values of the observed cost more often than higher values. This in turn would make us expect to find lower values of the average of the cost values in the group/sample more often than higher values of the average cost. Now, if the members of the group/sample are drawn from the population independently of each other then we conclude that cost-minimizing behavior implies that the probability of observing a specific group/sample decreases as the observed average cost of the group/sample increases. In the terminology used in probability theory and statistics: cost-minimizing behavior implies that the likelihood of the sample is a decreasing function of the average cost of the sample. Note that the likelihood denotes the probability of obtaining a particular sample of size N of decision makers. We use the word "likelihood" to remind us that we are talking about the probability of the whole sample and not the probability of one individual decision maker.

This summarizes the approach taken in this book. We assume that the probability, the likelihood, of the sample is a decreasing function of the average cost in the sample. This assumption has far reaching consequences in that logit type models for individual behavior follow from this assumption. Also, logit type models satisfy this assumption, i.e., logit type models express cost-minimizing behavior. The assumption can be tested against observations.

Our approach is essentially the efficiency principle (Smith 1978a,b, 1983, 1988; Erlander 1985; Erlander and Smith 1990), but, as mentioned above, we have changed perspective. Here we use the idea of the efficiency principle to *define* cost-minimizing behavior, and from this definition we derive all models. This means that we rely heavily on work by TE Smith as will be seen in the Notes sections in the following.

More structured problems can also be treated. Let there be trips from certain origin zones to certain destination zones. There is a choice between private car, bus and train. This is a combined distribution and mode choice problem. It can be treated with our approach by using activity constraints as introduced by Smith (1978a). The resulting model is a logit type model, but it is different from the traditional nested logit model.

By a *logit type model* or, as we shall often say, a *logit model* we mean a discrete probability distribution p_k for the alternative choices $k = 1, \ldots, K$, where the logarithm of the probability, $\log p_k$, can be written as a linear expression in some (generalized) cost values c_k. The (generalized) cost may include different costs and times as well as values of attributes that characterize the choice situation or the decision maker. In other words, logit type models are log linear functions, linear in the sense that the logarithm of the choice probability $\log p_k$ is a linear function of the cost and attribute components.

1.3 Intuitive Gravity Models and Most Probable State Approach

"Man tends of necessity to gravitate towards his fellow-man. Of all animals he is the most gregarious, and the greater the number collected in a given space the greater is the attractive force there exerted, as is seen to have been the case with the great cities of the ancient world, Nineveh and Babylon, Athen and Rome, and as is now seen in regard to Paris and

London, Vienna and Naples, Philadelphia, New York and Boston. Gravitation is here, as everywhere else in the material world, in the direct ratio of the mass, and in the inverse one of the distance." Carey (1858)

Many studies in social science and regional science deal with interaction between individuals or groups of individuals. One early work was Carey (1858) on immigration cited above. He was probably the first one to state the idea underlying the gravity model, i.e. that the number of people moving is proportional to the attractive forces and inversely proportional to the distance. Similar ideas were used by Ravenstein (1985) in his study of migration. Carey and Ravenstein presented their arguments in prose without using any formulae. The direct analogue with Newton's classical gravity formula requires the exponent of the distance in the denominator to be two. This was also the form obtained in the pioneering work by Lill (1891) on railway travel. In his study of migration Kulldorff (1955) found that an exponential distance factor, hence a logit type model, gave better agreement with migration data.

Borrowing from statistical mechanics Wilson (1967) derived the gravity formula for the most probable number of trips from zone i to zone j,

$$T_{ij} = \exp(\alpha_i + \beta_j - \gamma c_{ij}),$$

by maximizing the number of micro states subject to marginal constraints and a total (system) cost constraint. This is clearly a logit type formula since the exponent of the exponential expression is a linear function in the cost values c_{ij} with coefficients α_i, β_j and γ. Erlander (1977) showed that the same result is obtained by minimizing total (system) cost subject to an entropy constraint.

Wilson's problem can be described in the following way. Let there be given the number of trips originating in each origin zone and the number of trips terminating in each destination zone. Then the number of micro states resulting in the trip table $[T_{ij}]$ is $T!/\prod_{ij}(T_{ij}!)$. This combinatorial expression is approximately equivalent to the entropy $\sum_{ij} T_{ij} \log T_{ij}$. The most probable trip matrix is then obtained as a logit model by maximizing the combinatorial formula assuming that all micro states are equally probable and adding a constraint on the system cost $\sum_{ij} c_{ij} T_{ij}$.

This approach does not, however, explain the process through which the trip makers cooperate to satisfy the constraint on system cost. The same criticism can be directed against Erlander's derivation of the gravity model by interchanging cost and entropy as constraint and objective function of the optimizing problem. However, Erlander does not pretend that his formulation gives an explanation of the gravity model as a description of traveler's behavior. Instead he discusses efficiency and accessibility as means of evaluating different solutions of the model. In this respect the entropy can be seen as a measure of the degree of freedom of choice (Erlander and Schéele 1974; Erlander 1977).

Wilson (1970) developed a whole family of models from his maximum entropy principle, which was implicit in Wilson (1967).

The intuitive approaches to the gravity formulations as well as the maximum entropy approach or, equivalently, the most probable state approach lead to solutions of logit type. However, these approaches do not explain what kind of interaction between trip makers leads to the fulfillment of the marginal constraints as well as the total cost constraint. We shall see that this interaction can be seen as a consequence of the new probabilistic description of the cost-minimizing behavior of the trip makers used in this book.

Kruithof (1937) introduced the balancing procedure for computing the solutions to the gravity model applied to telephone traffic. The convergence of the balancing procedure was proved by Bregman (1967) and Evans (1970).

The maximum entropy formalism dealt with the most probable macro state given the marginal sums in the trip table and given total cost. This was further studied by TE Smith, who developed the most-probable-state analysis as a general analytical framework for testing a wide range of spatial models Smith (1985, 1990). For a review of the development see Roy and Thill (2004). A complete treatment of the gravity model was given by Sen and Smith (1995). See also Erlander and Stewart (1990).

1.4 User Equilibrium in a Network

> "The journey times on all the routes actually used are equal, and less than those which would be experienced by a single vehicle on any unused route." Wardrop (1952)

This is the celebrated Wardrop's first criterion. It defines deterministic *user equilibrium* in a network in the sense that no driver can reduce his journey time by choosing a new route. Beckmann showed how the equilibrium conditions can be derived by solving an optimization problem with the Beckmann integral $\sum_l \int_0^{v_l} c_l(v)\mathrm{d}v$ as objective function (Beckmann et al. 1956). A comprehensive discussion of Beckmann's contribution is given by Boyce (2007).

The probabilistic counterpart is the *stochastic user equilibrium* given by the simple (multinomial) logit model $p_k = \exp(-\gamma c_k)/\sum_{k=1}^{K} \exp(-\gamma c_k)$, $k = 1, \ldots, K$ (Daganzo and Sheffi 1977). Here the probability p_k denotes the probability that a random trip maker will choose the route k. Also for the stochastic user equilibrium the optimum conditions can be obtained by solving an optimization problem (Fisk 1980). Damberg et al. (1996) gave an algorithm for this problem.

1.5 Econometric Models of Probabilistic Choice

> "The relevance of these methods to economic analysis can be indicated by a list of the consumer choice problems, to which conditional logit analysis has been applied: choice of college attended, choice of occupation, labor force participation, choice of geographical location and migration, choice of number of children, housing choice, choice of number and brand of automobiles owned, choice of shopping travel mode and destination." McFadden (1974)

The paper from which the citation above is taken deals with the particular problem of choice of shopping travel mode and destination besides its pioneering development of the relevant theory. McFadden[1] continued developing theory and applying the theory to empirical investigations. In this way he studied, e.g., the work trip mode choice model for the Bay Area Rapid Transit BART with data collected in 1972 before BART was operational (McFadden 1978b). Later on McFadden (1987) investigated the housing status of singly elderly men on data from an investigation for Albany-Schectady-Troy, N.Y. given by Boersch-Supan and Pitkin (1982).

1.6 Luce's Axiomatic Derivation

> "One large portion of psychology – including at least the topics of sensation, motivation, simple selective learning, and reaction time – has a common theme: choice. To be sure, in the study of sensation the choices are among stimuli, in learning they are among responses, and in motivation, among alternatives having different preference evaluations; and some psychologists hold that these distinctions, at least the one between stimulus and response, are basic to an understanding of behavior. This book attempts a partial mathematical description of individual choice behavior in which the distinction is not made except in the language used in different interpretations of the theory. Thus the more neutral word "alternative" is used to include the several cases." Luce (1959)

Luce (1959) was the first to derive axiomatically a discrete choice model of logit type for the individual decision making with constant utilities. In this way the simple (multinomial) logit model was obtained. The choice probability distribution for the (multinomial) logit model is given by $p_k = \exp(-\gamma c_k)/\sum_{k=1}^{K}\exp(-\gamma c_k)$, $k = 1, \ldots, K$, (Luce and Suppes 1965).

1.7 ARUM – Additive Random Utility Maximization – Approach

> "The classical economically rational consumer will choose a residential location by weighing the attributes of each available alternative – accessibility of workplace, shopping, and schools; quality of neighborhood life and the availability of public services; costs, including housing price, taxes, and travel costs; dwelling characteristics, such as age, number of rooms, type of appliances; and so forth – and picking the alternative which maximizes utility." McFadden (1978a)

The standard approach in deriving choice probability functions is the ARUM, Additive Random Utility Maximizing, approach developed by Ben-Akiva (1974), Williams (1977), McFadden (1974, 1978a, 1981) and others.

The classical approach in deriving logit models postulates utility maximizing behavior. It is assumed that the behavior of a randomly chosen decision maker from the given population of decision makers can be described by a probability distribution p over the choice set. The standard derivation assumes that the

[1] McFadden shared the Prize in Economic Sciences in memory of Alfred Nobel in 2000.

perceived utility of alternative k is composed of the sum of the payoff v_k and a random non observable component X_k. The decision maker is assumed to maximize his or her perceived utility and thus choosing the alternative that maximizes $v_k + X_k$. From the additional assumption that the random variables $X_1, \ldots, X_K$ are independent and extreme value distributed follows the logit probabilities, $p_k = \exp(\theta v_k)/\sum_{k=1}^{K} \exp(\theta v_k)$. This is the (additive) random utility maximizing approach, ARUM. It has been generalized in various directions.

1.8 Structured or Nested Logit Models

> "The problems of analytic structure, with which this paper is concerned, relate to the mathematical representation of the trip decision process and, in particular, the form of the demand function and the relationship between the socioeconomic, land-use, and transportation-system variables which enter it." Williams (1977)

In the simple logit model all alternatives are treated equally. The only thing that matters is the (generalized) cost of the alternative. There are, however, many situations where a hierarchical or tree structure is appropriate. E.g., in modeling choice of travel mode and destination it may be argued that there are hidden differences between modes not covered by costs alone. In this case we would like to compare costs within each mode. This can be done by using structured models. There are many ways of formulating structured models.

Structured models are constructed for two different cases, "revealed" choices or "stated choices". The first case refers to a situation where the trip makers have taken a particular alternative, whereas the second case deals with an experimental situation where the trip makers are confronted with a number of hypothetical alternatives and make a choice based on descriptions of the costs and other attributes.

In spite of the long development of structured models beginning with Ben-Akiva (1974), Williams (1977) and McFadden (1978a) there is a continuing discussion on the correct choice of model and estimation of the parameters (see e.g. Koppelman and Wen 1998; Hensher and Greene 2002; Bliemer et al. 2009).

This book offers a new way of formulating and deriving structured models.

1.9 Transportation Problem in Linear Programming

> "Suppose that m warehouses (origins) contain various amounts of a commodity which must be allocated to n cities (destinations). Specifically, the ith warehouse must dispose of exactly the quantity a_i , while the jth city must receive exactly the quantity b_j." "The number c_{ij} represents the cost of shipping a unit quantity from origin i to destination j. Our problem is to determine the number of units to be shipped from i to j in order that stockpiles will be depleted and needs satisfied at an over-all minimum cost." Dantzig (1963)

The citation above describes the classical transportation problem in linear programming. Monge (1781) was the first to formulate this problem. He was motivated by the need to move masses of soil in reconstructing a military fortress. Monge was a mathematician who was raised to the office of Minister of the Marine under Bonaparte. When Napoleon fell so did Monge.

Kantorovich (1939) and Koopmans (1951) developed the theory around the transportation problem and later on linear programming in general.[2] Dantzig discovered the Simplex method which underlies most algorithms for solving these problems.

There is a close relationship between the gravity model for trip distribution and the transportation problem in linear programming (Evans 1973). The latter is a limiting case of the gravity model if the cost parameter goes to infinity. In this case the over-all cost goes to its minimum value.

1.10 Lagrangian Methods of Deriving Logit Models

> "Recent advances in network equilibrium modeling provide efficient algorithms for solving the urban trip assignment problem. These models can be extended to incorporate the trip distribution problem with two types of variable demand functions. By reinterpreting the zone-to-zone trip variable, these models can be viewed as urban location models." Boyce (1980)

Many different logit models can be derived by what can be called Lagrangian methods. By combining the Beckmann integral (Beckmann et al. 1956) with various entropy expressions and network constraints an optimization problem can be written in Lagrangian form with solutions characterized by user equilibrium in the network. Many people have contributed to this development (Murchland 1966; Tomlin 1967; Florian et al. 1975; Evans 1976; Erlander 1977; Boyce 1980; Boyce 1984). Boyce et al. (1988) give an overview including the combined trip distribution and assignment problem (Evans 1976; Erlander and Schéele 1974; Erlander 1977) and extensions, e.g., to mode choice problems. Logit models have also been used for models of urban location (e.g. Boyce 1980; Mattsson 1984; Abrahamsson and Lundqvist 1999).

Characteristic of these models is the formulation of a mathematical optimization problem having solutions with desirable properties such as user equilibrium assignment to the network and trip distribution of gravity type. The theoretical solution of the mathematical problem not only results in solutions of the desired type. In applications it can be used to compute numerical solutions as well. The latter is important. In fact, it may be quite useful to use one derivation of the model to give better understanding and motivation of the model and then use a mathematical optimization problem to produce numerical solutions in applications.

Lagrangian approaches are used to analyze problems in spatial interaction which contain logit models. We give one example regarding public transit (Schéele 1977, 1980). In this formulation waiting times depend inversely on the frequencies of

[2] Kantorovich and Koopmans shared the Prize in Economic Sciences in memory of Alfred Nobel in 1975.

buses thus causing non linear cost functions. The resulting optimization problem – compound minimization problem – is solved numerically. The trip matrix has gravity structure but the route choices do not.

1.11 Welfare Measures

> "Insofar as choosing is itself valuable, the existence and extent of choice have significance beyond that of providing only the means of choosing the particular alternative that happens to be chosen.... The foundational importance of freedom may well be the most far-reaching substantive problem neglected in standard economics." Sen (1988)

Average cost or expected cost can be used in evaluating different solutions in a planning situation. Welfare measures are more sophisticated. The importance of freedom of choice was stressed by Sen (1988, 1991, 1995).[3] We shall propose a welfare measure based on expected cost and expected freedom of choice.

In classic discrete choice theory, ARUM theory, composite utility – the so called log sum – has the interpretation of expected achieved perceived utility. In microeconomics it therefore has the interpretation of indirect utility. Williams (1977) identified the composite cost interpretation as the consumer surplus function and noted that it can take negative values. This fact was considered by Fisk and Boyce (1984) and others to be inconsistent with its interpretation as expected achieved cost within the ARUM approach. Composite utility/cost is widely used.

We shall see that our measure of advantage is equal to composite utility, thus giving a new interpretation of this quantity (Erlander 2005).

[3] Amartya Sen was awarded the Prize in Economic Sciences in memory of Alfred Nobel in 1998.

Chapter 2
Empirical and Policy Relevance of the New Paradigm

Abstract This chapter discusses the empirical and policy relevance of the new paradigm.

2.1 Empirical Relevance

Taking cost-minimizing behavior as defined in this book as starting point for model building is very natural. In fact, in dealing with rational decision makers, it is hard to think of a behavior where cost minimizing would not be an element of the decision making. Indeed, the models derived by the Lagrangian paradigm satisfy the cost-minimizing behavior criteria. This holds for any formulation where the objective function contains entropy terms. One example is the case of the most probable flows in Proposition 32, below, obtained as the optimal solution to a Lagrangian formulation. The most probable flows are given in gravity model form (8.16). Hence they satisfy the criteria for cost-minimizing behavior (Sect. 5.5).

Models derived by the Additive Random Utility Maximization paradigm, ARUM, also satisfy the cost-minimizing behavior criteria. This is shown in Chap. 4, with the most general case given in Proposition 5.

Models derived by the Lagrangian paradigm as well as models derived by the ARUM paradigm satisfy the cost-minimizing criteria. The situation is shown in the following diagram:

$$\text{Lagrangian} \Longrightarrow \text{Cost-Minimizing Behavior} \Longleftarrow \text{ARUM}$$

It is not entirely a matter of taste which paradigm is chosen for the derivation of the models. The paradigm chosen should produce relevant formulas, but perhaps even more important – it should give insight and understanding. The cost-minimizing paradigm gives a deeper understanding of what is at work in the models. There are also other differences. The Lagrangian formulation offers a way to calculate numerical solutions. This can be useful irrespective of the way the models are derived.

S. Erlander, *Cost-Minimizing Choice Behavior in Transportation Planning*,
Advances in Spatial Science, DOI 10.1007/978-3-642-11911-8_2,

The *derived* formulas can be tested against observations irrespective of which paradigm is used. However the cost-minimizing paradigm is the only approach which offers a way to test the underlying structure of the models. In this case the cost-minimizing criteria can be the object of empirical investigations (Chap. 7).

In the case of the ARUM paradigm the underlying assumptions cannot be observed, since the approach is based upon formulations in terms of unobservable stochastic variables (Axioms 1–5 in Sect. 4.5). Unobservable random variables, obviously, cannot be observed. There is also some doubt about the justification of the use of Gumbel distributed random variables, "No convincing justification has been advanced for the assumption of Gumbel distributed utilities, other than that it yields a particularly tractable model, namely the logit model" (Bell and Iida 1997, p. 122).

Nested or structured models are commonly used where there is an interest in grouping some of the choice alternatives together, e.g. when modeling mode choice behavior. However, the standard nested logit model should not be looked upon as a model of behavior. The hierarchical structure of the mathematical formulas invites a stepwise interpretation of the decision process. The hierarchical structure should be looked upon as a conceptual structure in a situation "where we might wish to regard the choices contained in a particular subset of the total choice set as having more in common with each other than with the remaining choices" (Daly and Zachary 1978). In our formulation of the structured logit model (Definition 12 and Proposition 10) activity equivalence is used to group the decision alternatives. Our approach has behavioral content and brings out exactly what mechanism is at work. Our approach gives new structured logit models. In order to obtain the standard nested logit model from cost-minimizing behavior we have to include comparisons of composite cost (log sum, inclusive values) in a peculiar way (Definition 14). This is a further indication that the standard nested logit model should not be looked upon as a behavioral model.

2.2 Policy Relevance

Models are used to study and analyze traffic systems. In a planning context it is important to predict future traffic flows and how they are distributed on different modes of travel and on the road network. In particular it is essential to analyze the effects of construction of new arterials and roads and changes in cost structure, e.g., by the introduction of toll roads.

Even more important is the analyses of welfare aspects of the transportation system. In this book, inspired by the work by Sen (1988, 1991, 1995), we introduce a welfare measure including cost and freedom of choice. This welfare measure is based on the idea that the value of choice depends on the costs of the alternatives but also that there is an intrinsic value in the possibility of choice. Our welfare measure turns out to be equal to the standard measure used in the ARUM paradigm, namely the composite utility (log sum, inclusive value). Thus we have a new interpretation of composite cost. This also explains some anomalities in the behavior of the composite cost for limiting values of the logit parameter.

However, there is a case for a warning at this point. There is a temptation to introduce too many variables. The models may seem complicated. Nevertheless they take care of only a few characteristics of the transportation system. In principle the models can be constructed to include many variables such as attributes of the trip maker (age, sex, income, . . .). This essentially means dividing the population of trip makers into groups. The choice probabilities are exponential functions of variables and parameters. Peculiar to this use of the exponential function is that there is a tendency to give more weight to a factor which is divided into many groups. For example painting buses red and blue has the effect of giving public transport more weight.

Chapter 3
Behavioral Foundations of Spatial Interaction Models

Abstract This chapter gives an overview of the content of the book. The book deals with a new approach to logit type discrete choice probability models – for transportation networks in particular. The models are derived from a new definition of cost-minimizing behavior – the likelihood of a sample is decreasing as a function of average cost. The formal definitions are given in Chap. 4: Definition 1 (multinomial logit model) and Definition 4 (general logit model). The results for the multinomial logit model and the general logit model are obtained in Propositions 1 and 3 respectively. All logit type choice probability functions satisfy the new definition. The new definition is in Part I applied to networks with constant link costs. It is shown that the simple (multinomial) logit model exhibits cost-minimizing behavior. Furthermore cost-minimizing behavior implies the logit model. A number of standard logit models are derived – stochastic route choice model, multi-attribute discrete choice model, gravity model and the general logit model. New structured logit models, different from the standard nested logit model, are obtained. A welfare measure based on cost and a measure of freedom of choice is given. The new welfare measure is shown to be identical with composite cost. The presence of cost-minimizing behavior in an observed data set can be investigated by using the property that the likelihood is decreasing as a function of average cost. This is used in constructing a graphical test for cost-minimizing behavior. In Part II of the book the new definition of cost-minimizing behavior is extended to the case of volume dependent separable link costs. Here equilibrium is studied. Cost-minimizing behavior implies that the most probable trip patterns are user equilibria. The most probable flow patterns are approximately obtained by solving the optimization problem obtained by relaxing the integer constraints and replacing the cumulative cost function with the Beckmann integral. Models are derived for route choice, combined choice of origin, destination and route as well as combined choice of origin, destination, mode and route.

S. Erlander, *Cost-Minimizing Choice Behavior in Transportation Planning*,
Advances in Spatial Science, DOI 10.1007/978-3-642-11911-8_3,

3.1 Basic Ideas – Cost-Minimizing Behavior and Equilibrium

Modern cities are very complex systems. The transportation and localization patterns are the result of a large number of decisions. Some of these decisions are made by community authorities and city planners. Other decisions are made by companies and institutions. The vast number of decisions, however, are made by citizens who choose where to live, where to work and how to make the work trip, to mention a few of the decisions. Many decisions are short term and deal with how the existing infrastructure can be used. Medium term decisions deal with where to live and where to work. Other decisions are long term and aim at improving the infrastructure in terms of the street network, bridges and tunnels, transit by trains and buses or influencing the users by changes in taxes or the price system. Hence, at the city planning level there is a need for predicting how the system will react to changes, and to evaluate the result. This can be accomplished by using models that attempt to describe various aspects of the system. The object of this book is the study of mathematical models for the behavior of travelers that can serve as instruments at the city planning level for predicting and evaluating the traffic system.

Transportation Planning and Location Science build complex sophisticated models that are widely used in practice. Many of the models contain some common elements. The starting point for this book is to try to find a few fundamental principles that govern behavior, and then to see how far these fundamental principles can take us into the world of different models.

The planner's selection of a mathematical model is a delicate question. It is important that the model can be validated and tested against data. It is also desirable to have a model where the impact of changes in various factors can be understood and analyzed.

Computers can analyze large amounts of data. Hence it might be argued that it would be sufficient for the city planner to collect as much data as possible and then use statistical methods for analyzing the data. However, this analysis has to be supported by a theory for the behavior of the travelers. The safest weather prediction is sometimes the statement that tomorrow's weather will be the same as today's. However, this is useless in a situation of changing weather. A sophisticated theory is needed when there is the possibility of change. Similarly, to predict future traffic we need a theory for the behavior of the system. This theory of behavior for the system has to start with a theory of behavior for the travelers.

The book focuses on fundamental principles underlying the models. The idea is to show how some new formulations of the elementary behavioral assumptions can form the basis for better understanding the structure and implications of various models. As examples of what can be obtained by our approach some standard models are derived. Also one or two new models are given. It is to be hoped that this new presentation of some well known standard models will stimulate and promote research into new models and different extensions of existing models.

Within the chosen framework many problems in Transportation and Location Science can be analyzed, including the route choice problem, the mode choice problem, the trip distribution problem, and combinations of these problems.

We shall take as our starting point two basic ideas, or principles, governing the behavior of the travelers. The *first principle* goes back at least to Adam Smith. In the traffic system context it may be formulated: other things being equal, travelers will choose the routes with shortest travel time or minimum cost, the least costly mode for travel and similarly for the choice of destination. The travelers are rational; they exhibit *cost-minimizing behavior*. The *second principle* or *criteria* was formulated by John Wardrop: network *equilibrium* obtains if in a network all used routes from one point to another point have equal travel time, and if no unused route has a lower travel time. The two principles are deterministic principles. We shall reformulate them in probabilistic terms in order to be able to treat systems containing large numbers of decision makers. This will also lead to a deeper understanding of the behavior of the traffic system.

By *cost* we shall mean the "out-of-pocket costs" directly experienced by the trip maker as a result of his trip or generalized cost. Generalized cost is often used in transportation analysis. It includes important factors such as travel time and direct out-of-pocket costs but excludes costs inflicted on the community that are not directly experienced by the trip maker as a result of his trip. Generalized cost is often represented by a linear function of factors believed to be important to the decision making. Our derivation of the multi-attribute model gives a new background for the notion of generalized cost.

In Part I we shall investigate the first principle, cost-minimizing behavior for networks with constant link costs. Equilibrium in networks with volume dependent separable link costs will be studied in Part II.

The path that we have chosen in exploring the basic principle is slightly different from standard approaches used in Transportation Science. However, most of the classical results will be obtained and some new light will be thrown on some of the commonplace models and their properties. Hence we believe that some new understanding can be brought about by investigating the consequences for transportation network modeling of formulating in our way the hypothesis that people behave according to the basic principles.

The probabilistic counterparts of the principles above will be the basis of our analyses. We shall investigate the consequences for the behavior of the system and how the resulting state of the system may be evaluated.

Logit models are often used in modeling discrete decision problems encountered in transportation studies, but also in many other fields. The models that we are going to study turn out to be variants of logit models. The models are characterized by the functional form for the probability of choosing each specific alternative. This probability represents the choice behavior of one individual decision maker/traveler chosen randomly from the population of decision makers. The quality of the model, i.e., the extent to which the model can mirror observed choices and can be used for predicting future situations depends on the specification of functional form for the choice probability function and on how well the underlying assumptions represent the choice behavior observed in practice.

Our main focus will be on decisions taken by travelers in a network. However, we shall start by investigating the choice problem in general without the restriction to travelers in a network. From a new formulation of the first principle of elementary behavior – cost-minimizing behavior – we shall derive the functional form of the probability of choice. The functional form of the probability function can be recognized as the standard (multinomial) logit model.

We shall make precise what should be meant by cost-minimizing behavior and derive from this the choice probability distribution for the simple (multinomial) logit model. The general case is then examined. Derivations of some particular models used in Transportation Science follow before discussing freedom of choice and benefit measures. Our treatment of freedom of choice and benefit measures will lead to a new interpretation of the standard composite cost/log sum measure. We give a graphical test for investigating if the decision makers behave according to our definition of cost-minimizing behavior. The graphical test is applied to simulated data. Estimation of parameters is briefly commented upon in the Appendix (Chap. 9).

Cost-minimizing behavior is defined in Part II for separable flow dependent link costs. Conditions for equilibrium are investigated. The most probable trip patterns are found to be user equilibrium trip patterns. Conditions for most probable flow patterns are given in terms of an optimization problem. Relaxing the integer constraints, approximate solutions can be identified as standard formulations for problems where the number of trips is treated as a continuous variable. The approach used in this book gives a better motivation for the optimization formulations and also gives a deeper understanding of how traffic systems work.

The particular models given are simple examples of what can be obtained by the approach used. The particular models can be combined in many ways into more complex models covering many different situations. We shall, with a few exceptions, refrain from going into these more complex models since we wish to present the basic ideas rather then giving a total overview of all models that can be derived.

The book is organized in the following way. References are, with a few exceptions, not given in the main text. At the end of each chapter is a Comments and Extensions section as well as a Notes section. The latter gives some background and brief historical notes as well as references to books and papers used in the text.

To facilitate reading and to make the chapters to some extent self contained there is some repetition.

3.2 Probability Models

Why probability models? There are two reasons for using probability models in the present context. Firstly, in traffic and transportation systems the number of decision makers – the individual trip makers – is large. It is not possible to handle each decision maker separately. Rather, when discussing the behavior of decision makers we shall treat them as independent samples drawn from a large population of decision makers, thus describing and analyzing the behavior by means of a probability model. Secondly, another consequence of the large number of decision makers is that in the final analysis we want simple measures describing various features of the traffic system in order to evaluate different solutions. It would be impossible to keep track of every specific decision maker. This means that our main interest will lie in deriving average or expected measures.

In using a probability model we cannot say what decision a specific decision maker will reach. We can only give the probability. We shall use the probability of choice in the following way. The probability of choice will be the probability of choosing one specific alternative amongst the available choice alternatives for a *randomly* chosen decision maker.

Probability models are used to cover differences and variations that cannot be ascribed fully to variation in variables, or to attributes used to characterize the decision situation and the properties of the decision makers. Probability models aim at taking care of the stochastic variation. However, there is no guarantee that the model is true in the sense that it describes and explains the real world satisfactorily. Hence, careful model selection is crucial for the model to be useful. A poor model may produce numbers but will be of little help in analyzing data.

Let there be $k = 1, \ldots, K$ choice alternatives for the decision maker and let the probability of choosing alternative k be p_k.

Now there are several paths open to us. We can make the a priori assumption that the probability function p_k has some given form, e.g., the logit model form that we shall later discuss.

More satisfactory would be to make some intuitively motivated assumptions about cost-minimizing behavior and then to derive the form of the choice probability function from these assumptions. This is the approach used in the classic Additive Random Utility Maximizing, ARUM, approach (McFadden 2001). The details of this approach will be given in Sect. 4.5.1. The assumptions of the ARUM approach have to be assumed a priori, since the utilities cannot be observed. The basic assumptions cannot be tested, only derived properties can be tested. We shall return to this discussion in Sect. 4.7. When the form of the choice probability function p_k has been established, observations on behavior are used to estimate the parameters of the probability function p_k by taking independent samples from the population of decision makers.

We shall go one step further in extracting information from independent samples of observations. Recent development in resampling techniques has shown the great potential of this approach in model selection. In particular, computer intensive statistical methods using resampling can be of great help in model selection. We shall not use resampling in deriving the form of the choice probability function p_k. However, inspired by the new interest and progress in drawing conclusions directly from independent samples, we shall extract information about the exact form of the probability distribution including possible parameters by formulating the conditions for minimizing behavior in terms of properties of *independent samples of observations*. Surprisingly enough, this approach will make it possible to derive the exact form of the probability distribution p_k (for one randomly chosen decision maker). Note that our approach is conceptual or theoretical. The conditions for minimizing behavior will be assumed to hold for every independent sample. From this assumption follows, as we shall see, the mathematical form of the choice probability function.

However, observed samples are necessary for estimating parameters and for testing whether the assumptions are satisfied (Chap. 7).

The approach used in this book is different from the classic Additive Random Utility Maximizing approach. The underlying idea is to use information contained in independent samples of decision makers to find the form of the choice probabilities. The classical approach obtains the form of the probability distribution by assuming a priori a detailed specification of the decision procedure including non observable random utilities. The elements of the specification cannot be tested since the random utilities cannot be observed. Independent samples are then used to estimate parameters of the choice probability distribution. We go one step further. We draw conclusions about the *form* of the choice probability distribution from testable assumptions regarding the behavior of cost minimizing decision makers in independent samples of decision makers. The assumptions are given in a form that makes it possible to test the assumptions. Hence the assumptions themselves, not only the consequences of the assumptions, can be refuted by observations. This is not the case in the classical approach.

3.3 Freedom of Choice

Average or expected cost is important in evaluating decision making. Another quality of interest is related to the dispersion of the decisions. We shall call this *freedom of choice*. We have chosen the term "freedom of choice" to indicate the richness of the opportunities open to the decision makers. This is a often neglected quality of the decision making. We shall return to this in the Notes section of Chap. 6.

By "freedom of choice" can be meant many things. In our case we characterize the decision making by a probability distribution over the available alternatives. This means that we do not describe the decision making for a specific decision maker. Central to our analyses is the probability of choice for a randomly chosen decision maker out of the population of decision makers. Thus freedom of choice has to be defined accordingly. Freedom of choice is related to how constrained the decision making is or how dispersed the decisions are. Since the decision making in our approach is governed by a choice probability distribution, the constraints are implicit and show up in the variance or dispersion of the probability distribution. Great variance/dispersion corresponds to great freedom of choice as we shall now explain.

We shall introduce a quantity which can be viewed as a *measure of revealed freedom of choice*. This quantity is based on the values obtained in an independent sample. The underlying idea is that the observed frequencies, as well as the choice probability distribution, reveal something about how constrained the decision making is. If the decisions are uniformly distributed over the choice alternatives, then there are no constraints and the level of freedom of choice is high. The dispersion is also high. On the other hand, if the decisions concentrate on a smaller set of the alternatives, then the dispersion and the level of freedom of choice is lower. In this case the choices are constrained by one factor or another such as the cost of alternatives or physical barriers.

Consider a sample $d = (d_1, \ldots, d_N)$ of decisions. This sample gives rise to observed frequencies $z = (z_1, \ldots, z_K)$. Here z_k denotes the number of occurrences of alternative k in the sequence of decisions $d = (d_1, \ldots, d_N)$. There will usually be many samples $d = (d_1, \ldots, d_N)$ giving the same frequencies $(z_1, \ldots, z_K)$. Note that $\sum_{k=1}^{K} z_k = N$. The number of possible outcomes of each sample d is given by the combinatorial formula $N!/(z_1! \cdots z_K!)$.

This number can be used as a measure of revealed freedom of choice. If the observed frequencies $(z_1, \ldots, z_K)$ can be obtained in many ways then there are few constraints in the decision making. On the other hand, if there are few ways to generate the observed frequencies, then the decision making is more constrained. Consider two extreme cases. First let $z_1 = N$ and $z_k = 0$ for $k = 2, \ldots, K$. In this sample every decision maker has chosen alternative 1. There is only one way of obtaining this sample. This means that there is no freedom of choice, every decision maker has chosen alternative 1. The reason for these choices may be that the alternatives $k = 2, \ldots, K$ are extremely costly or even impossible due to physical barriers. Second let $z_k = N/K = z$, an integer, $k = 1, \ldots, K$. This means that the decisions are uniformly spread out over the alternatives. There is a high degree of freedom of choice, since there are many samples $d = (d_1, \ldots, d_N)$ giving the frequencies $z_k = N/K = z$, $k = 1, \ldots, K$. The observed frequencies can be obtained in $N!/(z! \cdots z!) = N!/(z!)^K$ ways. This number may be very large, thus indicating a high degree of freedom of choice or dispersion.

Thus the combinatorial formula can be used as a measure of revealed freedom of choice. The measure of revealed freedom of choice will be combined with a measure of achievement based on average negative cost into a measure of advantage or welfare (Chap. 6).

3.4 Cost-Minimizing Behavior

The essential question is, how would cost-minimizing behavior show up in some property of independent samples?

As before, let p_k be the probability of choosing alternative k and let c_k be the corresponding cost. Let z_k, $k = 1, \ldots, K$, denote the number of times alternative k is chosen in a random independent sample $d = (d_1, \ldots, d_N)$ of N decision makers. The probability of obtaining d is $\Pr(d) = \Pr(d_1) \cdots \Pr(d_N) = \prod_{k=1}^{K} p_k^{z_k}$. This expression is in statistical theory known as the *likelihood* $\mathcal{L}(z)$ of the sample, $\mathcal{L}(z) = \prod_{k=1}^{K} p_k^{z_k}$. The likelihood is here a function of the z frequencies only. To stress this fact we prefer using the likelihood $\mathcal{L}(z)$ rather than the identical probability $\Pr(d) = \Pr(d_1) \cdots \Pr(d_N)$. In the following when we talk about the likelihood of a sample it must be borne in mind that the likelihood $\mathcal{L}(z)$ is equal to the probability of a particular sequence of decisions d, $\Pr(d) = \Pr(d_1) \cdots \Pr(d_N) = \mathcal{L}(z)$. (In order to obtain the probability of the frequencies we have to multiply by the number of ways to obtain z, $N!/(z_1! \cdots z_K!)$. Hence the probability of z is $\Pr(z) = (N!/(z_1! \cdots z_K!))\mathcal{L}(z)$. This probability will however not be used in the following.)

The basic assumption about cost-minimizing behavior is formulated intuitively for independent random *samples* of decision makers in terms of the likelihood (probability) of the particular choices made, and the corresponding (average) cost of the sample, $\bar{c}(z) = \sum_{k=1}^{K} c_k(z_k/N)$ in the following way:

- *Cost-minimizing behavior is said to prevail if samples with lower average cost* $\bar{c}(z) = \sum_{k=1}^{K} c_k(z_k/N)$ *are more frequent, i.e., if the likelihood* $\mathcal{L}(z) = \prod_{k=1}^{K} p_k^{z_k}$ *of the sample* $d = (d_1, \ldots, d_N)$ *is a decreasing function of the average cost* $\bar{c}(z)$.

From this basic assumption about cost-minimizing behavior the choice probability function can be derived. It turns out to be the same logit type probability distribution as is obtained in the classic Additive Random Utility Maximizing derivation. In our case all random variables can be observed and thus the basic assumption about cost-minimizing behavior can be investigated and tested. This way of formulating criteria for cost-minimizing behavior can be extended to more elaborate situations. Most standard logit type models can be derived from this approach.

It is important to notice that our approach formulates criteria for cost-minimizing behavior, which the unknown choice probability function must satisfy. We then derive the form of the choice probability function under the assumption that these criteria are satisfied (Sect. 4.2).

3.5 The Simple (Multinomial) Logit Model Exhibits Cost-Minimizing Behavior

In this section we shall illustrate the implications of our approach for the simplest case: the simple (multinomial) logit model.

Let p_k denote the probability of choosing alternative k with cost c_k. The probability distribution for the logit model is given by

$$p_k = \frac{\exp(-\gamma c_k)}{\sum_{k=1}^{K} \exp(-\gamma c_k)},$$

where γ is a parameter to be estimated.

As can be seen from the formula the logit model has the property that the probability of choosing a particular alternative k is a decreasing function of the cost c_k associated with the alternative. Hence alternatives with a high value of the cost have a small probability of being chosen.

For the logit model the property that the choice probability function p_k is decreasing in the cost c_k has its counterpart for the likelihood. When observing a sample of decision makers, the likelihood (probability) of observing a particular sample of decisions is a decreasing function of the average cost of the sample, as will be seen now.

Consider an independent sample of size N of decision makers. Let z_k be the number of decision makers choosing alternative k. The likelihood of observing a sample $d = (d_1, \ldots, d_N)$ with frequencies z where $z = (z_1, \ldots, z_K)$ is given by

$$\mathcal{L}(z) = \prod_{k=1}^{K} p_k^{z_k}.$$

By using the logit probability given above this becomes

$$\mathcal{L}(z) = \prod_{k=1}^{K} (\frac{\exp(-\gamma c_k)}{\sum_{k=1}^{K} \exp(-\gamma c_k)})^{z_k}.$$

By taking logarithms we obtain

$$\log \mathcal{L}(z) = -\gamma \sum_{k=1}^{K} c_k z_k - N \log \sum_{k=1}^{K} \exp(-\gamma c_k) = -\gamma N \bar{c}(z) - N \log \sum_{k=1}^{K} \exp(-\gamma c_k),$$

where $\bar{c}(z)$ is the average cost $\bar{c}(z) = (1/N) \sum_{k=1}^{K} c_k z_k$. Hence for the logit model given above the likelihood of observing $(z_1, \ldots, z_K)$ is a decreasing function of the average cost $\bar{c}(z)$. Hence the logit model satisfies our definition of cost-minimizing behavior. Lower values of the average cost will be more frequently observed and the logit model is compatible with cost-minimizing behavior.

Similar results hold for all logit type probability functions.

3.6 Cost-Minimizing Behavior Implies the Logit Model

The argument can be reversed, starting with the definition of cost-minimizing behavior and from this definition deriving the form of the probability distribution. Note that the choice probability function is unknown at the outset. The definition of cost-minimizing behavior was given before:

- *Cost-minimizing behavior is said to prevail if observations with lower average cost* $\bar{c}(z) = \sum_{k=1}^{K} c_k(z_k/N)$ *are more frequent, i.e., if the likelihood* $\mathcal{L}(z) = \prod_{k=1}^{K} p_k^{z_k}$ *of the sample* $d = (d_1, \ldots, d_N)$ *is a decreasing function of the average cost* $\bar{c}(z)$.

  The main result will be obtained in Sect. 4.2:

- *Cost-minimizing behavior implies the logit model:*

$$p_k = \frac{\exp(-\gamma c_k)}{\sum_{k=1}^{K} \exp(-\gamma c_k)}.$$

According to the discussion above the logit model exhibits cost-minimizing behavior. The remarkable fact is that the logit model is the only choice probability function that exhibits cost-minimizing behavior in terms of average cost. Hence cost-minimizing behavior implies the logit model for the choice probabilities. The same holds for all logit type – log linear – choice probability functions provided that cost-minimizing behavior is defined correspondingly. The general case is treated in Sect. 4.3.

The relation between the logit model and cost-minimizing behavior is expressed in the following implication:

$$\text{Cost-Minimizing Behavior} \Longleftrightarrow \text{Logit Model}$$

3.7 Welfare Measure

The frequencies $(z_1, \ldots, z_K)$ can be obtained in $N!/\prod_{k=1}^{K} z_k!$ ways. This combinatorial expression is approximately equivalent to the observed entropy $H([z_k/N]) = -\sum_{k=1}^{K}(z_k/N)\log(z_k/N)$. The latter will be taken as a measure of revealed freedom of choice. For the logit model the observed entropy divided by the parameter γ of the logit model turns out to have the dimensional property of cost. A welfare measure can then be obtained by combining minus the average cost and observed entropy (expressed in cost terms by dividing by the parameter γ of the logit model) and taking expectations

$$W = E[-\bar{c}(z) + \frac{1}{\gamma}H([z_k/N])] \approx -\frac{1}{\gamma}\log\sum_{k=1}^{K}\exp(-\gamma c_k),$$

which is the standard composite cost (Chap. 6). This is a new *welfare measure* including cost and freedom of choice. Hence in our approach the composite cost has the interpretation of the sum of expected negative cost and expected freedom of choice. The standard interpretation is that the composite utility represents *expected achieved perceived utility*, i.e., the indirect utility.

3.8 Graphical Test

The presence of cost-minimizing behavior in an observed data set can be investigated by using the property that the likelihood is decreasing as a function of the average cost. This will be used in constructing a graphical test for cost-minimizing behavior (Chap. 7). Let there be J independent samples $z^j = (z_1^j, \ldots, z_K^j)$, $j = 1, \ldots, J$. Then the observed likelihood or the negative of the observed entropy $-H([z_k^j/N])$ can be plotted against the observed average cost $\sum_{k=1}^{K} c_k z_k^j/N$ for

$j = 1, \ldots, J$. Cost-minimizing behavior will then appear as a clustering of the points around a line with slope $-\gamma$. If enough observed samples are not available, resampling techniques can be used to generate the J samples from one original sample. The properties of the graphical test will be studied in Chap. 7 by means of simulation.

3.9 Some Particular Discrete Choice Models

The simple logit model can be extended in two ways, by introducing (a) multiple cost values and (b) activity constraints. This will be briefly discussed now.

So far we have introduced one cost value c_k for each alternative k. The cost structure is one dimensional. In general however there will be many cost measures, such as monetary costs of various kind and time "costs", connected with each alternative. If there are S cost values c_{sk}, $s = 1, \ldots, S$ for each alternative k, then the cost structure is S-dimensional.

In the simple case above there is one constraint on the observed frequencies, $\sum_{k=1}^{K} z_k = N$, leading to one parameter γ in the logit model. We often wish to classify the alternatives, e.g., according to choice of mode of transportation. This can be done by introducing activity constraints. The activity constraints will reduce comparisons to comparisons within each group. When dealing with mode choice we wish to restrict attention to activity equivalent samples, i.e., to samples where the number of car trips and the number of public transport trips respectively are equal. By introducing multidimensional cost structures and activity constraints on the observed frequencies various standard logit type models can be derived. Amongst the models derived are the simple (multinomial) logit model, the general logit model, (Chap. 4), stochastic route choice model, multi-attribute discrete choice model, gravity model for trip distribution, structured (nested) logit models, standard nested logit model and models with socioeconomic factors (Chap. 5).

3.10 Equilibrium

In the second part of the book we study *equilibrium for systems with separable flow dependent link costs*. The results obtained in Part I carry over to this more general situation. Cost-minimizing behavior implies that the most probable trip patterns are user equilibria in the sense that no trip maker can choose a less costly trip for the trip from an origin to a destination. However, there is one difference from Wardrop's conditions: every admissible trip pattern u will have positive probability to occur, even if this probability is very small for most trip patterns. Therefore this kind of equilibrium is called "dispersed equilibrium".

The results obtained are surprisingly strong. The standard models are usually derived from a mathematical optimization problem where for instance an entropy term gives a gravity model type solution while at the same time a Beckmann integral produces user equilibrium route choices. The motivation for this approach is that the formulated mathematical problem leads to the desired structure of the solutions. Instead, in this book, cost-minimizing behavior implies the structure. This gives a deeper understanding of what is at work, providing as it does an explanation based on behavior.

The essential step in studying equilibrium with flow dependent separable link costs is the replacement of constant link costs with Smith's cumulative cost function.

Let there be T trip makers, each choosing a specific route $r_t, t = 1, \dots, T$. A *trip pattern* u contains T routes, $u = (r_1, \dots, r_T)$. Let $c(r_t|r_1, \dots, r_{t-1})$ be the cost to trip maker t if he chooses route r_t when the first trip makers have chosen routes $(r_1, \dots, r_{t-1})$. Let $c(u)$ denote the *cumulative user cost function* defined by

$$c(u) = \sum_{t=1}^{T} c(r_t|r_1, \dots, r_{t-1}).$$

If the behavior of the trip makers is cost minimizing, then the last trip maker will choose a route r_T that minimizes the last term of the cumulative cost function $c(r_T|r_1, \dots, r_{T-1})$. One of the important properties of the cumulative cost function is that the value of the cumulative cost function $c(u)$ is the same irrespective of the order of the trips in the trip pattern u. This means that every trip maker t may be considered as the last one. Hence, if cost-minimizing behavior prevails, we would expect the cumulative cost function $c(u)$ to take lower values more frequently than higher values. Indeed,

- $c(u^*) = \min c(u)$ *implies that* u^* *is a user equilibrium.*

Let there be one independent sequence of trip patterns $u^1, \dots, u^N$, and let z_k be the number of times the particular trip pattern u_k appears in the sequence.

Let p_k be the probability of obtaining trip pattern u_k and let $c(u_k)$ be the corresponding value of the cumulative cost function.

The definition of cost-minimizing behavior becomes:

- *Cost-minimizing behavior is said to prevail if observations with lower average cumulative cost* $\sum_{k=1}^{K} c(u_k)(z_k/N)$ *are more frequent, i.e., if the likelihood* $\mathcal{L}(z) = \prod_{k=1}^{K} p_k^{z_k}$ *is a decreasing function of the average cumulative cost* $\sum_{k=1}^{K} c(u_k)(z_k/N)$.

The main result for route choice will be obtained in Sect. 8.3:

- *Cost-minimizing behavior for route choice implies the logit model:*

$$p_k = \frac{\exp(-\gamma c(u_k))}{\sum_{k=1}^{K} \exp(-\gamma c(u_k))}.$$

From this formula it follows that the most probable trip patterns are obtained when the cumulative cost function is at a minimum, which in turn implies that the most probable trip patterns are user equilibria. Hence

- *For flow dependent separable link cost, let cost-minimizing behavior prevail. Then the most probable trip patterns are user equilibria.*

3.11 Choice of Origin, Destination and Route

The previous results can be extended in many ways. We give the combined distribution and route assignment problem. This problem is also extended to include mode choice.

Let $v_{ijr}(u)$ = the number of trips in the trip pattern u from origin $i = 1, \ldots, I$ to destination $j = 1, \ldots, J$ using route r. We define

$$T_{ij}(u) = \sum_r v_{ijr}(u), \quad A_i(u) = \sum_j T_{ij}(u), \quad B_j(u) = \sum_i T_{ij}(u).$$

$A_i(u)$ and $B_j(u)$ can be identified as the marginal totals of the trip matrix $[T_{ij}(u)]$.

- *Cost-minimizing behavior implies the gravity type combined distribution and assignment model–CDA model:*

$$p_k = \exp[\sum_i \alpha_i A_i(u_k) + \sum_j \beta_j B_j(u_k) - \gamma c(u_k)].$$

Note that this model is a dispersed model formulated in integer variables. It is different from the standard CDA model. Let $a(u)$ denote the vector of all marginal totals,

$$a(u) = (A_1(u), \ldots, A_I(u), \ B_1(u), \ldots, B_J(u)).$$

Keeping this activity vector constant, $a(u) = a$, we find that the most probable trip patterns are obtained when the cumulative cost function is at a minimum, which in turn implies that the most probable trip patterns are user equilibria.

- *For flow dependent separable link cost, let cost-minimizing behavior prevail. Then, in the CDA model, the most probable trip patterns are user equilibria.*

Many trip patterns correspond to the same flow pattern $[v_{ijr}]$. The number of trip patterns generating the flow pattern $[v_{ijr}]$ is given by the combinatorial formula $T!/\prod_{i,j,r}(v_{ijr}!)$. The probability of the flow pattern $[v_{ijr}]$ is obtained by multiplying the probability p_k with the combinatorial formula, and the most probable flow patterns are given as the optimal solution to problem 8.12 below. Let v_l denote the flow on link l generated by the flow pattern $[v_{ijr}]$, and let $c_l(k)$ be the cost on link l when the flow on the link is k. Then the cumulative cost function can be written $c(u) = \sum_l \sum_{k=1}^{v_l} c_l(k)$. Then follows the most important result of Sect. 8.4.2:

- *Cost-minimizing behavior implies that the most probable flow patterns are given as the optimal solutions to the problem*

$$\max \quad \frac{T!}{\prod_{i,j,r}(v_{ijr}!)} \exp[\sum_{i,j,r} \alpha_i v_{ijr} + \sum_{i,j,r} \beta_j v_{ijr} - \gamma \sum_l \sum_{k=1}^{v_l} c_l(k)],$$

subject to

$$\sum_{j,r} v_{ijr} = A_i,\ i = 1, \ldots, I,\ \sum_{i,r} v_{ijr} = B_j,\ j = 1, \ldots, J.$$

This is a remarkable result and gives a deeper understanding of how the traffic system works. However, in practical applications the restriction to integer values of the variables make the computations unnecessarily complicated. The theoretical exact results have corresponding continuous deterministic approximate versions. These approximate versions are the same as the optimizing approaches formulated in the standard derivations. In order to obtain the continuous counterparts we shall relax the restriction to integer values and let the variables be real numbers and replace the relevant summations of the link costs (the cumulative cost function) with Riemann integrals in order to obtain what we here call Beckmann integrals.

The main results are the following (Proposition 32):

- *Let the trip pattern distribution represent cost-minimizing behavior. Then the most probable flow patterns given the marginal totals are approximately given by the unique optimal solution $v^* = [v^*_{ijr}]$ to the problem*

$$\min_{i,j,r} \quad [\sum_{i,j,r} v_{ijr}(\log v_{ijr} - 1) + \gamma \sum_l \int_0^{v_l} c_l(v) dv],$$

subject to $v_{ijr} \geq 0$, for all i, j, r and

$$\sum_{j,r} v_{ijr} = A_i,\ i = 1, \ldots, I,\ \sum_{i,r} v_{ijr} = B_j,\ j = 1, \ldots, J.$$

Furthermore the optimal solution satisfies for some real numbers α_i and β_j

$$v^*_{ijr} = \exp(\alpha_i + \beta_j - \gamma c^*_{ijr}),$$

where

$$c^*_{ijr} = \sum_l c_l(v^*_l)\delta_{ij,lr}.$$

If cost-minimizing behavior prevails for the individual trip makers then we approximate the most probable flows by solving the optimization problem above. The solution has gravity model structure for the flows and user equilibrium for the route choices. This optimization problem is the same as the problem used in the standard approach but here we obtain it from cost-minimizing behavior.

3.12 Choice of Origin, Destination, Mode and Route

The previous results are extended to the case of simultaneous choice of origin, destination, mode and route.

3.13 Comments

The standard model of microeconomics deals with competitive equilibrium in a market where production and demand meet and prices and quantities are determined by the intersection of supply and demand curves. The variation between different decision makers is taken care of by the supply and demand functions. Decision makers are entering or leaving the market depending on prevailing prices. The welfare or benefit is measured by consumer surplus defined as the sum of the areas behind the demand curves if marginal utility of income is constant.

Discrete choice models, on the other hand, model the variation between decision makers by choice probability functions describing cost-minimizing behavior. Traditionally the so called composite cost is used to evaluate benefits. We use a slightly different approach, separating advantage and achievement and including freedom of choice. This gives a new interpretation of composite cost.

3.14 Notes

The new approach used in this book was first published by Erlander and Lundgren (2004). It represents a shift in perspective from efficient population behavior to cost-minimizing behavior. Many propositions and proofs are reformulations of results presented in Smith (1978b, 1983, 1988), Erlander (1985) and Erlander and Smith (1990).

We give a few examples of standard textbooks: Wilson et al. (1981), Ben-Akiva and Lerman (1985), Sheffi (1985), Erlander and Stewart (1990), Anderson et al. (1992), Patriksson (1994), Sen and Smith (1995), Bell and Iida (1997), Hensher and Button (2000). Overviews of the development of the Random Utility Maximization approach are given by McFadden (2001) and Boyce and Williams (2005).

3.15 About Notation

Greek letters $\alpha, \beta, \gamma, \ldots$ will be used to denote parameters used in the models. They will often have the interpretation of (Lagrange) multipliers connected with constraints and cost expressions. It is difficult to uphold general rules for when to use one or the other of the Greek letters. In most cases though, the letters β and γ will be related to cost expressions. When dealing with the standard Additive Random Utility Maximizing approach, the letter θ will be used.

Part I
Cost-Minimizing Behavior: Constant Link Costs

We shall in Part I focus on the first principle: cost-minimizing behavior. We shall be concerned with the choice probability formulation in a network with volume independent link costs and then, in Part II, proceed to the case with volume dependent link costs.

Chapter 4
Logit Models for Discrete Choice

Abstract The ideas introduced in Chap. 3 are made more precise. The form of the choice probability function is derived from basic assumptions about cost-minimizing behavior. The simple (multinomial) logit model as well as discrete choice logit models in general are treated. Axiomatic derivations are used to discuss and compare our approach and the standard ARUM formulation.

4.1 Preliminaries

In many planning situations the concern is to estimate the number of decision makers making each choice among a set of available discrete alternatives. Models describing this type of choice are known as discrete choice models. Typical applications concern land-use and travel demand analysis, including residential and employment location and the choice of frequency, origin, destination and mode of travel. We shall use probability models to characterize decision behavior. The probability distribution takes care of variations in the decisions due to randomness or to factors unknown to us. These factors may very well be known to the decision maker, and hence his/her decision may be deterministic, but we choose to model them by including them together with other factors in the random variation ascribed to the probability distribution.

The aim in making the notion of cost-minimizing behavior more precise is to derive from basic principles an expression for the probability of choosing each alternative. Then, we can easily compute the expected proportion of all decision makers (or the expected total number) choosing each alternative.

We shall now make more precise the ideas introduced in Chap. 3.

We are going to derive the form of the choice probability function from basic assumptions about cost-minimizing behavior.

For each choice alternative we specify a cost, which incorporates all relevant characteristics of the alternative. This cost is composed by known deterministic quantities for each alternative, the same for all decision makers under consideration. (The case with varying circumstances of the decision makers will be treated in Sects. 5.8 and 5.9.) In a choice situation between different modes of travel the

S. Erlander, *Cost-Minimizing Choice Behavior in Transportation Planning*, Advances in Spatial Science, DOI 10.1007/978-3-642-11911-8_4,

cost can be the simple travel cost, or a cost composed by monetary costs, time costs, parking costs and so on. In Sect. 4.2 we concentrate on the simple case where the cost of each alternative is given by a single real number. In Sects. 5.3 and 5.4 we investigate the case where the cost is specified as a linear function of the variables (attributes) used to characterize the alternatives. We shall also see how the notion of activity equivalence can be used to handle constraints and socio-economic differences between the decision makers.

In what follows the word "cost" will for each alternative denote a single cost or a generalized cost representing cost and time costs and other attributes. By cost we mean in the simplest case out of pocket costs arising directly to the decision maker/trip maker as a consequence of choosing a particular alternative. Perceived costs, or rather, perceived utilities are used in the ARUM approach.

The decisions of the individuals in the population can be different, and alternatives with higher as well as lower cost can be chosen. A reasonable condition on the probability distribution describing these choices is that an alternative with lower cost would be more probable. In other words, our intuitive feeling for the notion of cost-minimizing behavior requires that the probability p_k of choosing alternative k must be a non increasing function of the cost c_k. This is not enough, however, to characterize completely cost-minimizing behavior. There are many non increasing functions that would be candidates for the probability p_k. In order to determine the probability function p_k we shall define cost-minimizing behavior in terms of properties of a sample of decision makers.

We said above that for cost-minimizing behavior for one decision maker to obtain the probability p_k must be non increasing in the cost c_k of the alternative k. This intuitively motivated property for one decision maker can be extended to a sample of N decision makers by considering the total sum of the costs for the members of the sample. Consider a sample of decision makers. The individuals will choose different alternatives; also alternatives with a higher cost will be chosen by some individuals in the sample. However, the probability for a (randomly chosen) decision maker in the sample to choose an alternative with higher cost is lower. This means that we would expect an alternative with lower cost to be chosen more often, i.e., such an alternative would be more frequently observed. Our assumption of cost-minimizing behavior for the sample then says that cost-minimizing behavior obtains if lower values of *the sum of the costs* for the chosen alternatives in the sample are more probable – more frequently observed. It turns out that by using this simple device we can derive the desired probability distribution. Thus, by assuming cost-minimizing behavior in this way we can derive the choice probabilities without further assumptions about the specific behavior of the individual decision makers.

In Chap. 3 cost-minimizing behavior was defined in terms of average cost for the sample. From now on we shall use instead the sum of the costs, i.e., the total cost. The definition of cost-minimizing behavior is also slightly changed so as to suit the proofs. The content of the definition is the same.

In Sect. 4.2 we describe and define the notion of cost minimization behavior applied to the most simple case of a discrete choice situation. The same ideas are treated axiomatically in Sect. 4.4. In Sect. 4.3, we present cost-minimizing behavior in the general case of discrete choice modeling. The general ideas are then applied to some commonly used models in Chap. 5. In particular, we show in Sects. 5.3 and 5.5 how the multi-attribute multinomial logit model and the gravity model for trip distribution, respectively, can be derived from our definition. The standard Additive Random Utility Maximizing approach is given in Sect. 4.5.1.

4.2 The Simple (Multinomial) Logit Model

Let Ω be the set of all decision makers under consideration. The set Ω may be trip makers in a city making decisions about where to live and where to work, commuters choosing between public transit or private car or car drivers deciding which route to take in a network. Each member of the set Ω chooses in some way an alternative k from the discrete set of choice alternatives $k = 1, \ldots, K$ with corresponding cost c_k, $k = 1, \ldots, K$. According to the first principle the decision makers are rational; they exhibit cost-minimizing behavior. This would imply that alternatives with lower cost values would be chosen more frequently. By drawing at random one decision maker from the population Ω of decision makers we can observe which alternative k with cost c_k that has been chosen by the decision maker. This observation alone does not give any information about presence or absence of cost-minimizing behavior. We need more than one observation to be able to make comparisons. To draw conclusions about cost-minimizing behavior we shall take a random independent sample of size N of decision makers. Let d_n, $n = 1, \ldots, N$, be the decision taken by decision maker n in the sample and let *decision pattern* $d = (d_1, \ldots, d_N)$ denote the decisions of all N individuals in the sample. Note that the individual decisions d_n as well as the vector d of all decisions by the decision makers in the sample are stochastic variables.

Now the essential question is, how would cost-minimizing behavior show up in the sample?

Let $p_k = \Pr(d_n = k)$ denote the probability that a decision maker drawn at random from the population Ω chooses alternative k, $k = 1, \ldots, K$. The decision of each individual can be described by the unknown discrete probability distribution $p = (p_1, \ldots, p_K)$. This is the probability distribution we want to derive and it is the same for all individuals in the sample as well as in the population Ω.

We wish to determine the form of the probability distribution p corresponding to cost-minimizing behavior. It is intuitively satisfying to require that p_k is non increasing in c_k. However, this is not enough to determine the form of p_k. Can more be obtained by studying *samples* of decision makers? Indeed, the form of the probability distribution can be determined in this way as we shall demonstrate in the following.

The costs incurred by the N decision makers can be written $(c_{d_1}, \ldots, c_{d_N})$. If the behavior of the decision makers is cost minimizing, we expect lower values of c_{d_n} to be more frequent. Let $z_k(d)$ denote the number of times that alternative k is chosen in decision pattern d. Furthermore, let $z(d) = (z_1(d), \ldots, z_K(d))$. The cost

$c(d)$ of the decision pattern d is defined by

$$c(d) = c_{d_1} + \cdots + c_{d_N} = \sum_{k=1}^{K} c_k z_k(d). \tag{4.1}$$

Assume that the decision makers choose alternatives independently. The probability of decision pattern d can be expressed as the likelihood

$$\Pr(d) = p(d_1) \cdots p(d_N) = \prod_{k=1}^{K} p_k^{z_k(d)},$$

where $p(d_n)$ is the probability of the alternative chosen by individual n, provided that the sample is an independent sample. The likelihood is a function of the z-values only, so that we can write the likelihood

$$\mathcal{L}(z) = \prod_{k=1}^{K} p_k^{z_k(d)}.$$

For later use we shall define average cost, expected cost and composite cost. Average cost is defined by

$$\bar{c}(d) = \frac{1}{N} \sum_{k=1}^{K} c_k z_k(d).$$

Expected cost is defined by

$$E[\bar{c}(d)] = \sum_{k=1}^{K} c_k p_k.$$

Composite cost is defined by

$$\tilde{c} = -\frac{1}{\gamma} \log \sum_{k=1}^{K} e^{-\gamma c_k}. \tag{4.2}$$

4.2.1 *Formal Derivation of the Simple (Multinomial) Logit Model*

We shall now derive the functional form of the choice probabilities p_k for the simpliest case: the (multinomial) logit model. We shall formulate the definition of cost-minimizing behavior in terms of total cost rather than average cost as we did in Chap. 3. To make the definition operational we shall formulate the definition by comparing two independent samples. The content of the definition below is that cost-minimizing behavior is obtained if and only if the likelihood of a sample of decision makers is a non increasing function of the total cost of the sample.

Consider two independent random samples of the same size N and compare the two decision patterns, d^1 and d^2, for the samples. Let $z_k^1 = z_k(d^1)$ and $z_k^2 = z_k(d^2)$ denote the corresponding number of times alternative k is chosen in decision pattern (sample) 1 and 2, respectively.

Definition 1 (Cost-Minimizing Behavior). A probability distribution $p=(p_1, \ldots, p_K)$ represents cost-minimizing behavior if and only if, for any sample size N, and for any two decision patterns d^1 and d^2,

$$c(d^1) \leq c(d^2) \Longrightarrow \Pr(d^1) \geq \Pr(d^2),$$

or equivalently

$$\sum_{k=1}^{K} c_k z_k^1 \leq \sum_{k=1}^{K} c_k z_k^2 \Longrightarrow \prod_{k=1}^{K} p_k^{z_k^1} \geq \prod_{k=1}^{K} p_k^{z_k^2}. \tag{4.3}$$

We shall say that a probability distribution p is a *cost-minimizing* probability distribution if it satisfies condition (4.3).

Note that this way of defining cost-minimizing behavior is *conceptual*. Condition (4.3) must hold for any two samples of the same size N and for any sample size N. The condition compares the likelihood of the two samples.

Condition (4.3) may seem rather mild. However, it is sufficient for deriving the form of the probability distribution p. The factor at work here, as can be seen from the proof of Proposition 1 below, is the fact that (4.3) must hold for any value of N.

Note that cost-minimizing behavior is here defined in terms of functions of an independent sample of decision makers, namely the functions $c(d)$ and $\Pr(d)$. Thus cost-minimizing behavior is formulated in terms of observations on behavior. (The classical derivation of the choice probability function defines cost-minimizing behavior by specifying the choice procedure in terms of unobservable/hidden utilities. See Sect. 4.5.1.)

This way of defining cost-minimizing behavior is very natural. *In fact it is hard to imagine a reasonable definition of cost-minimizing behavior that would not imply this.* This definition of cost-minimizing behavior does not contradict other definitions. In particular, all models that can be derived by the additive random utility maximizing approach with Generalized Extreme Value (GEV) distribution for the non observable random components satisfy our definition (see Sect. 4.5.1).

The remarkable fact is that the functional form of the probabilities p_k can be derived from cost-minimizing behavior according to the formulation above.

Definition 1 is expressed in terms of properties of decision patterns involving simultaneous choices by samples of decision makers. It will, however, be used to derive the probability distribution p for *one* decision maker. The simultaneous probabilities in (4.3) are, because of independence, expressed in terms of the probability distribution p which is assumed to hold identically for each member of the sample of decision makers. The right hand side of (4.3) expresses the relation between the probabilities of the particular decision patterns, and a decision pattern with lower total cost is in Definition 1 assumed to be more probable than a decision pattern with higher total cost.

Note that if p is a cost-minimizing probability distribution, then, taking $N = 1$, it follows from the definition above that p_k is non increasing in c_k for $k = 1, \ldots, K$.

Given that we assume cost-minimizing behavior to hold, the form of the probability distribution can be derived. A remarkable fact is that this simple definition (Definition 1) is enough to completely determine the probability distribution p.

The following proposition gives necessary and sufficient conditions for cost minimizing behavior in the simplest case.

Proposition 1 (Cost-Minimizing Behavior in the (Multinomial) Logit Model). *The probability distribution $p = (p_1, \ldots, p_K)$ represents cost-minimizing behavior in the sense of Definition 1 if and only if it is a log linear probability distribution $p_k = \exp(\mu - \gamma c_k)$ which can equivalently be written*

$$p_k = \frac{\exp(-\gamma c_k)}{\sum_{k=1}^{K} \exp(-\gamma c_k)}. \tag{4.4}$$

Proof. Outline of proof. The idea is to start from any integer valued strictly positive vector $\tilde{z}$ which corresponds to decision pattern frequencies $\tilde{z} = (\tilde{z}_1, \ldots, \tilde{z}_K)$ and then try to find another vector z, such that z and $\tilde{z}$ together satisfy the left hand side of inequality (4.3). This will be realized by solving the linear program (4.5) below with objective function $\sum_{k=1}^{K} z_k \log p_k$ to be minimized and constraint $\sum_{k=1}^{K} c_k z_k \leq \sum_{k=1}^{K} c_k \tilde{z}_k$. Assuming that the costs are rational numbers there is a rational optimal solution $\hat{z}$. Multiplying with the smallest common denominator n the vector $n\hat{z}$ has integer values and corresponds to a decision pattern of size nN. Likewise the vector $n\tilde{z}$ corresponds to a decision pattern of size nN. This means that we have two decision patterns of the same size satisfying the left hand side of (4.3). From the assumption that p is cost minimizing then follows

$$\prod_{k=1}^{K} p_k^{n\hat{z}_k} \geq \prod_{k=1}^{K} p_k^{n\tilde{z}_k},$$

which implies

$$\sum_{k=1}^{K} \hat{z}_k \log p_k \geq \sum_{k=1}^{K} \tilde{z}_k \log p_k.$$

However, $\hat{z}$ is optimal in the linear program. Hence $\tilde{z}$ is also optimal. From dual complementarity follows that the dual constraints are satisfied with equality, $\log p_k = \mu - \gamma c_k$, since $\tilde{z}$ is strictly positive.

We shall now prove the proposition. By substitution into (4.3) it is easy to verify that the probability distribution (4.4) satisfies the definition of cost-minimizing behavior. To prove the reverse implication some more work is needed.

Assume that the probability distribution $p = (p_1, \ldots, p_K)$ is cost minimizing, hence satisfying (4.3). Let the costs c_k, $k = 1, \ldots, K$, be rational. Further, let $\tilde{z} = (\tilde{z}_1, \ldots, \tilde{z}_K)$ be arbitrary given, where $\tilde{z}_k > 0$ and integer, $k = 1, \ldots, K$, such that $\sum_{k=1}^{K} \tilde{z}_k = N$.

Consider the linear program in the K variables $z = (z_1, \ldots, z_K)$:

$$\min \sum_{k=1}^{K} z_k \log p_k,$$

$$\text{subject to} \sum_{k=1}^{K} c_k z_k \leq \sum_{k=1}^{K} c_k \tilde{z}_k, \ \sum_{k=1}^{K} z_k = N, \ z_k \geq 0, \ k = 1, \ldots, K \quad (4.5)$$

with dual

$$\max(\mu N + \gamma \sum_{k=1}^{K} (-c_k)\tilde{z}_k) \quad \text{subject to } \mu - \gamma c_k \leq \log p_k.$$

Clearly, $(z_1, \ldots, z_K) = (\tilde{z}_1, \ldots, \tilde{z}_K)$ is a feasible solution. We shall show that this solution is also optimal, and the form of the probability distribution then follows from duality theory.

Since the linear program has rational costs and rational right hand side, there is a rational optimal solution $\hat{z} = (m_1/n, \ldots, m_K/n)$ where m_k, $k = 1, \ldots, K$, and n are integer numbers. Now, consider two decision patterns defined by $d^1 = n\hat{z} = (n\hat{z}_1, \ldots, n\hat{z}_K)$ and $d^2 = n\tilde{z} = (n\tilde{z}_1, \ldots, n\tilde{z}_K)$. These decision patterns have the same sample size, which is

$$\sum_{k=1}^{K} n\hat{z}_k = \sum_{k=1}^{K} n\tilde{z}_k = nN.$$

Since $\hat{z}$ by construction is optimal in the linear program it is also feasible and we have that

$$n \sum_{k=1}^{K} c_k \hat{z}_k \leq n \sum_{k=1}^{K} c_k \tilde{z}_k,$$

which can be expressed

$$\sum_{k=1}^{K} c_k (n\hat{z}_k) \leq \sum_{k=1}^{K} c_k (n\tilde{z}_k).$$

This inequality shows that the decision patterns $d^1 = n\hat{z}$ and $d^2 = n\tilde{z}$ satisfy the inequality on the left hand side of implication (4.3) in Definition 1. Since p is assumed to be cost minimizing, it follows that

$$\prod_{k=1}^{K} p_k^{n\hat{z}_k} \geq \prod_{k=1}^{K} p_k^{n\tilde{z}_k},$$

which can be written

$$\sum_{k=1}^{K} \hat{z}_k \log p_k \geq \sum_{k=1}^{K} \tilde{z}_k \log p_k.$$

Hence, $z = \tilde{z}$ is also an optimal solution of the linear program (because $\hat{z}$ is optimal by assumption). Since $\tilde{z}_k > 0, \ k = 1, \ldots, K$, it follows from the complementary slackness theorem of linear programming that there are dual variables μ, and $\gamma > 0$, such that the dual constraints are satisfied with equality. Hence,

$$\log p_k = \mu - \gamma c_k,$$

which can be rewritten as the logit probability distribution (4.4). □

The proposition was proven assuming rational costs c_k. This is no restriction in applications. The proposition is true without this assumption but the proof is not given here.

We recognize the probability distribution (4.4) as the (multinomial) logit model. Hence, the logit model represents cost-minimizing behavior in the sense of Definition 1. Also, Proposition 1 says that the probability distribution (4.4) is the *only* model that has cost minimization behavior in the sense of Definition 1. Thus, we have demonstrated that any probability distribution that represents cost-minimizing behavior in the sense of formula (4.3) with rational costs has to be of the logit model form (4.4).

Substituting the costs c_k by payoffs (utilities) v_k changes the logit model to

$$p_k = \frac{\exp(\theta v_k)}{\sum_{k=1}^{K} \exp(\theta v_k)}, \quad k = 1, \ldots, K$$

and minimizing into maximizing. To remind ourselves that we have a maximizing formulation we have changed the parameter γ into θ. Let $v(d) = \sum_{k=1}^{K} v_k z_k$. We can then define payoff (utility) maximizing behavior analogous to Definition 1:

Definition 2 (Payoff (Utility) Maximizing Behavior). A probability distribution $p = (p_1, \ldots, p_K)$ represents payoff (utility) maximizing behavior if and only if, for any sample size N,

$$v(d^1) \geq v(d^2) \Longrightarrow p(d^1) \geq p(d^2),$$

or equivalently

$$\sum_{k=1}^{K} v_k z_k^1 \geq \sum_{k=1}^{K} v_k z_k^2 \Longrightarrow \prod_{k=1}^{K} p_k^{z_k^1} \geq \prod_{k=1}^{K} p_k^{z_k^2}. \tag{4.6}$$

We shall say that a probability distribution p is a *payoff (utility) maximizing probability distribution* if it satisfies condition (4.6) above.

Similarly we obtain the corresponding proposition:

Proposition 2 (Payoff (Utility) Maximizing Behavior in the (Multinomial) Logit Model). *The probability distribution $p = (p_1, \ldots, p_K)$ represents payoff (utility) maximizing behavior in the sense of Definition 2 if and only if it is a log linear probability distribution $p_k = \exp(\mu + \theta v_k)$, which can equivalently be written*

$$p_k = \frac{\exp(\theta v_k)}{\sum_{k=1}^{K} \exp(\theta v_k)}. \tag{4.7}$$

4.3 The General Logit Model for Cost-Minimizing Behavior

In the simple logit model treated in Sect. 4.2 there is one cost value, or generalized cost value, for each alternative in the choice set. The generalized cost may contain components such as monetary costs, time costs, parking costs etc. The generalized cost is obtained by adding the weighted components into one single cost measure. However, the components can also be treated as separate cost attributes, in which case the different weights will come out as coefficients. This will be treated in detail in Sects. 5.3 and 5.4. Also, in some situations when defining cost-minimizing behavior, we wish to restrict comparison to samples which are equivalent with regard to some aggregated measure of activity of the members of the sample. One example is the gravity model for trip distribution given in Sect. 5.5. In the trip distribution case it is natural to restrict attention to trip patterns which contain equal numbers of trips going from each origin zone and terminating at each destination zone.

The general case that we present now offers the possibility of formulating many different models.

In this section we shall generalize from the simple logit model, treated so far, to the general logit model. This means treating the case where we have several cost functions (cost measures) which influence the choice of the decision maker, and where the measure of activity of the sample is defined by a general set of linear equations. Cost-minimizing behavior is then defined with respect to all cost functions simultaneously, where we restrict the comparison to such decision patterns which are activity equivalent as defined by these linear equations. The main result is the Representation Theorem (Proposition 3), which shows that all cost-minimizing probability distributions are of exponential (log-linear) type. The results will be applied to a number of commonly used logit models in Chap. 5.

In the simple (multinomial) logit case (Definition 1) there was one implicit activity constraint, namely that the two decision patterns are of the same size N.

Everything in this section can be expressed in payoff or utility maximizing terms as well, by simply replacing cost with minus payoff or minus utility.

We shall start by repeating some of the notation used in Sect. 4.2.

We think of the decision makers as a population Ω of decision makers. Each decision maker makes one decision, i.e., he/she chooses one alternative from the set of choice alternatives.

To draw conclusions about cost-minimizing behavior we shall take independent random samples of size N of decision makers. Let d_n, $n = 1, \ldots, N$, be the decision taken by decision maker n in the sample and let *decision pattern* $d = (d_1, \ldots, d_N)$ denote the decisions of all N individuals in the sample. Note that the individual decisions d_n as well as the vector d of all decisions taken by the decision makers in the sample are stochastic variables.

Let $p_k = \Pr(d_n = k)$ denote the probability that a decision maker $n, n = 1, \ldots, N$, drawn at random from the population Ω chooses alternative k, $k = 1, \ldots, K$. The decision of each individual can be described by the unknown discrete probability distribution $p = (p_1, \ldots, p_K)$. This is the probability distribution we want to derive and it is the same for all individuals in the sample as well as in the population Ω.

The activity constraints as well as the cost functions will be formulated in terms of vectors and matrices. Only elementary vector algebra will be needed. In this section symbols for vectors and matrices are boldfaced.

Let $z_k(d)$ denote the number of times that alternative k is chosen in decision pattern d. Furthermore, let the vector of these frequencies be $\mathbf{z} = (z_1(d), \ldots, z_K(d))^T$.

Now, assume that we have S cost measures to consider and that activity equivalence is defined with respect to M activity measures. Then, we can introduce a *cost matrix* $\mathrm{C} \in \mathrm{R}^{S \times K}$, where element c_{sk} denotes the value of the cost function (cost measure) s for alternative k. Similarly, we introduce an *activity matrix* $\mathrm{A} \in \mathrm{R}^{M \times K}$, where element a_{mk} indicates the level of activity measure m in alternative k.

The activity matrix A will be used to define M linear constraints on the vector $\mathbf{z}$ of the form A$\mathbf{z}$, expressing the total level of the activity measures $1, \ldots, M$ in decision pattern d. We only want to compare decision patterns of the same size N and with identical levels of the activity measures. In most cases the entries in the A matrix will be 0 or 1. Hence, the expression A$\mathbf{z}$ essentially stands for summing over certain of the z_k-variables. The cost matrix C will be used to define S cost measures which we shall compare simultaneously.

Consider two independent decision patterns d^1 and d^2, with corresponding vectors $\mathbf{z}^1$ and $\mathbf{z}^2$, describing the choices of two samples of decision makers of the same size N.

Definition 3 (Activity Equivalence). For any given activity matrix $\mathrm{A} \in \mathrm{R}^{M \times K}$ two decision patterns of the same size N are Activity Equivalent if and only if

$$\mathrm{A}\mathbf{z}^1 = \mathrm{A}\mathbf{z}^2.$$

Note that the levels of the activity measures are not specified. The definition only requires the levels in the two decision patterns to be equal. However, the values of the parameters to be derived in Proposition 3 below will be determined by specifying the levels of the activity measures.

Note, furthermore, that since the z_k are integer values, the equality $\mathrm{A}\mathbf{z}^1 = \mathrm{A}\mathbf{z}^2$ will not hold for any arbitrary matrix A. In Proposition 3 below the matrix will be assumed to be rational.

We are now ready to define *Cost-Minimizing Behavior* in the general case.

Definition 4 (Cost-Minimizing Behavior in General). For any given activity matrix $\mathrm{A} \in \mathrm{R}^{M \times K}$ and cost matrix $\mathrm{C} \in \mathrm{R}^{S \times K}$, the probability distribution p is defined to be a cost minimizing probability distribution with respect to A and C if and only if for any independent decision patterns d^1 and d^2 of the same size N,

$$[\mathrm{A}\mathbf{z}^1 = \mathrm{A}\mathbf{z}^2, \mathrm{C}\mathbf{z}^1 \le \mathrm{C}\mathbf{z}^2] \quad \Longrightarrow \quad \prod_{k=1}^{K} p_k^{z_k^1} \ge \prod_{k=1}^{K} p_k^{z_k^2}, \tag{4.8}$$

holds for any N.

The definition contains M activity equivalence constraints and S cost measures. The inequalities $\mathbf{C}\mathbf{z}^1 \le \mathbf{C}\mathbf{z}^2$ must hold simultaneously for all S cost measures. Hence, a probability distribution exhibits cost-minimizing behavior if, when comparing activity equivalent decision patterns, smaller values of *all* the cost measures imply higher probability of the decision pattern to be observed. Nothing is said about other possible decision patterns, or about the probability distribution if not all the cost relations hold. We shall return to this discussion after Proposition 3.

Assuming a cost-minimizing probability distribution is equivalent to making the assumption that decision patterns with lower cost in all S cost components are at least as probable as decision patterns with higher costs when activities are equivalent.

Definition 4 covers more situations than appear at first sight. The alternatives corresponding to the alternatives $k = 1, \ldots, K$ can often be classified into sets with different activity constraints and/or different costs. As an example consider the trip distribution problem (gravity model treated in Sect. 5.5). In the trip distribution case the alternatives concern the simultaneous choice of origin i and destination j among a total of I origins and J destinations. Here T_{ij} denotes the number of trip makers going from i to j. The alternatives are now the pairs $(ij), i = 1, \ldots, I,\ j = 1, \ldots, J$ and $\mathbf{z}$ is replaced by $(T_{11}, \ldots, T_{1J}, T_{21}, \ldots, T_{2J}, \ldots, T_{I1}, \ldots, T_{IJ})^T$. We have $M = I + J$ activity measures, defining the total number of trip-makers going from some origin i to some destination j, respectively, and the matrix A is the standard transportation matrix. Thus, each row corresponds to one origin *or* one destination. Element a_{mk} will take the value of 1 if alternative k (relation (i, j)) concerns origin (destination) m. The general activity equations in Definition 3 are then defined according to (5.4), and the levels of activity are given by the marginal totals.

The definition of cost-minimizing behavior in general, Definition 4, is a natural generalization of the definitions used before with one cost condition (Definition 1). Note that in Definition 1 there is one implicit activity constraint, namely $\sum_{k=1}^{K} z_k = N$. Our definition for the general case makes it possible to derive the general form of cost-minimizing probability distributions by using results previously obtained for the efficiency principle.

From the assumption that a probability distribution is cost minimizing in the sense of Definition 4 follows the specific form of the probability distribution. This is given in the general Representation Theorem presented next.

Proposition 3 (Representation Theorem for Cost-Minimizing Probability Distributions). *For any given rational activity matrix* $\mathrm{A} \in \mathrm{R}^{M \times K}$ *and cost matrix* $\mathrm{C} \in \mathrm{R}^{S \times K}$, *the probability distribution* p *is a cost-minimizing probability distribution with respect to* A *and* C *if and only if there exists a nonnegative vector* $\gamma \in \mathrm{R}^{S}$, *such that for some vector* $\alpha \in \mathrm{R}^{M}$, *and some* $\mu \in \mathrm{R}$,

$$p_k = \exp(\mu + \alpha^T \mathbf{a}_k - \gamma^T \mathbf{c}_k), \; k = 1, \ldots, K, \tag{4.9}$$

where $\mathbf{a}_k$ *and* $\mathbf{c}_k$ *are column* k *of the matrices* A *and* C, *respectively.*

Here μ represents a scaling parameter that guarantees that the probabilities sum up to one. Formula (4.9) can be written

$$p_k = \frac{\exp(\alpha^T \mathbf{a}_k - \gamma^T \mathbf{c}_k)}{\sum_{k=1}^{K} \exp(\alpha^T \mathbf{a}_k - \gamma^T \mathbf{c}_k),}. \tag{4.10}$$

This is the form that will be used in most cases in the book.

Proof. The proof of the Representation Theorem is a straight forward generalization of the proof of Proposition 1.

By substitution into (4.8) it is easy to verify that the probability distribution (4.10) satisfies the definition of cost-minimizing behavior. To prove the reverse implication, again, some more work is needed.

Assume that the probability distribution $p = (p_1, \ldots, p_K)$ is cost minimizing, hence satisfying (4.8). Let the cost matrix C have rational coefficients c_{mk}, $m = 1, \ldots, M$, $k = 1, \ldots, K$. Further, let $\tilde{\mathbf{z}} = (\tilde{z}_1, \ldots, \tilde{z}_K)^T$ be arbitrary and given, where $\tilde{z}_k > 0$ and integer, $k = 1, \ldots, K$, such that $\sum_{k=1}^{K} \tilde{z}_k = N$.

Consider the linear program in the K variables $\mathbf{z} = (z_1, \ldots, z_K)^T$:

$$\min \sum_{k=1}^{K} z_k \log p_k, \;\; \text{subject to} \;\; \mathrm{A}\mathbf{z} = \mathrm{A}\tilde{\mathbf{z}}, \;\; \mathrm{C}\mathbf{z} \leq \mathrm{C}\tilde{\mathbf{z}}, \;\; \sum_{k=1}^{K} z_k = N,$$
$$z_k \geq 0, \; k = 1, \ldots, K.$$

By construction $\mathbf{z} = \tilde{\mathbf{z}}$ is a feasible solution of the linear program. It is also optimal, as we shall now show.

Since the linear program has rational coefficients and rational right hand side, there is a rational optimal solution $\hat{\mathbf{z}} = (m_1/n, \ldots, m_K/n)$ where $m_k,\ k = 1, \ldots, K$, and the smallest common denominator n are integer numbers. Now, consider two decision patterns defined by $d^1 : n\hat{\mathbf{z}} = (n\hat{\mathbf{z}}_1, \ldots, n\hat{\mathbf{z}}_K)$ and $d^2 : n\tilde{\mathbf{z}} = (n\tilde{\mathbf{z}}_1, \ldots, n\tilde{\mathbf{z}}_K)$. These decision patterns have the same sample size, which is

$$\sum_{k=1}^{K} n\hat{z}_k = \sum_{k=1}^{K} n\tilde{z}_k = nN.$$

Since $\hat{\mathbf{z}}$ by construction is optimal in the linear program it is also feasible and for $(n\hat{\mathbf{z}})$ and $(n\tilde{\mathbf{z}})$ we have

$$\mathrm{A}(n\hat{\mathbf{z}}) = \mathrm{A}(n\tilde{\mathbf{z}}) \text{ and } \mathrm{C}(n\hat{\mathbf{z}}) \leq \mathrm{C}(n\tilde{\mathbf{z}}).$$

It follows that the decision patterns $d^1 : n\hat{\mathbf{z}}$ and $d^2 : n\tilde{\mathbf{z}}$ satisfy the inequality on the left hand side of implication (4.8) in Definition 4. Since p is assumed to be cost minimizing, it follows that

$$\prod_{k=1}^{K} p_k^{n\hat{z}_k} \geq \prod_{k=1}^{K} p_k^{n\tilde{z}_k},$$

which can be written

$$\sum_{k=1}^{K} \hat{z}_k \log p_k \geq \sum_{k=1}^{K} \tilde{z}_k \log p_k.$$

Hence, $\mathbf{z} = \tilde{\mathbf{z}}$ is also an optimal solution of the linear program (because $\hat{\mathbf{z}}$ is optimal by assumption). Since $\tilde{z}_k > 0,\ k = 1, \ldots, K$, it follows from the complementary slackness theorem of linear programming that there are dual variables $\gamma \in \mathrm{R}^S$, such that, for some vector $\alpha \in \mathrm{R}^M$, and some $\mu \in \mathrm{R}$, the dual constraints are satisfied with equality. Hence,

$$\log p_k = \mu + \alpha^T \mathbf{a}_k - \gamma^T \mathbf{c}_k,$$

which can be rewritten as the logit probability distribution (4.10). □

The proposition was proven assuming that the cost matrix C has rational coefficients. This is usually no restriction in applications. The proposition is true without this assumption but the proof is not given here.

The Representation Theorem in its most general form is given in the Appendix, Theorem 1.

All standard discrete choice models of log linear type can be written in the form of (4.9), and are therefore cost minimizing in the sense of Definition 4. This means that they can be derived from cost-minimizing behavior by the "only if" implication of Proposition 3.

The result of Proposition 3 is a remarkable fact, since very little is assumed in order to obtain the specific form of the probability distribution. The reverse implication is also true: any probability distribution of the specific form is cost minimizing. Proposition 1 in Sect. 4.1 is a special case of Proposition 3 and the probability distributions (5.2) and (5.9) of the multi-attribute discrete choice model and the gravity model in Chap. 5 can also be derived using this proposition.

The principal strength of the definition of cost-minimizing behavior in Definition 4 is that very little has to be assumed about specific details of behavior in order to derive all cost minimizing probability distributions. Many different decision mechanisms are in principle compatible with the definition. The definition is expressed in observable quantities, and the assumptions can therefore be tested and refuted by means of observations. No assumption has to be made about the form of the probability distributions.

We shall now return to a discussion of the situation when condition (4.8) of the definition of cost-minimizing behavior, Definition 4, is not satisfied. Since decision patterns are stochastic there is of course no guarantee that condition (4.8) will be satisfied. This is not necessary either, since the definition assumes something only for certain decision patterns. As an example consider the discussion of how socio-economic factors can be handled in Sect. 5.9. Here activity equivalence is used in order to differentiate between income classes. In this case z_{ks} denotes the number of decision makers belonging to class s choosing alternative k with cost c_{ks}. Activity equivalence between two samples (z^1_{ks}) and (z^2_{ks}) is defined by the conditions $\sum_{k=1}^{K} z^1_{ks} = \sum_{k=1}^{K} z^2_{ks}$, $s = 1, \ldots, S$. There will be many decision patterns which do not satisfy these conditions. By restricting attention to decision patterns satisfying the conditions we compare only decision patterns with equal numbers in the income classes. Similarly, there are many decision patterns which violate the cost conditions. We simply do not assume anything about these decision patterns. It is enough to consider only a rather restricted set of decision patterns for the Representation Theorem above to hold.

4.4 Axiomatic Derivations of Logit Models

In order to bring out explicitly the difference between the hypothesis of cost-minimizing behavior used in this book and the hypothesis underlying the classical additive random utility maximization derivation we shall formulate the hypothesises axiomatically. We shall refer to Cost-Minimizing Behavior according to Definition 1 as the CMB approach. Payoff-Maximizing Behavior according to Definition 2 will be denoted by PMB. The treatment of cost-minimizing behavior and payoff-maximizing behavior is exactly parallel. The classical Additive Random Utility Maximization derivation will be denoted by ARUM and presented in Sect. 4.5

We shall alternate between the terms "payoff" and "utility". The reason for this is that the classical approach is strongly connected with the term "utility". In the rest of the book we have tried to uphold the distinction between "payoff" as denoting an observable quantity and "utility" as a quantity which can be observable or unobservable. Since we are mostly dealing with observable quantities the term "payoff" is usually used. In much of the economics literature "utility" denotes unobservable hidden quantities.

The material in Sects. 4.4 and 4.5 is not used elsewhere in the book.

4.4.1 Axioms for Cost-Minimizing Behavior

Let Ω be the set of all decision makers under consideration. The set Ω may be trip makers in a city making decisions about where to live and where to work, commuters choosing between public transit or private car or car drivers deciding which route to take in a network. Each member of the set Ω chooses in some way an alternative with corresponding (generalized) cost.

Let $p = (p_1, \dots, p_K)$ be the probability distribution of the decision taken by a decision maker drawn at random from the population Ω, and let c_k denote the cost of alternative k, $k = 1, \dots, K$. We wish to determine the form of the probability distribution p corresponding to cost-minimizing behavior. It is intuitively satisfying to require that p_k is decreasing in c_k. However, this is not enough to determine the form of p_k. Can more be obtained by studying *samples* of decision makers? Indeed, the form of the probability distribution can be determined in this way as we have seen in Sect. 4.2.

Let d_n denote the choice made by decision maker n. Consider an independent sample of size N of decision makers from the population Ω. Let the *decision pattern* $d = (d_1, \dots, d_N)$, denote the decisions taken by the N decision makers. Then the costs incurred by the N decision makers can be written $(c_{d_1}, \dots, c_{d_N})$. If the behavior of the decision makers is cost minimizing, we expect lower values of c_{d_n} to be more frequent. Let $z_k(d)$ denote the number of times that alternative k is chosen in decision pattern d. Furthermore, let $z(d) = (z_1(d), \dots, z_K(d))$.The cost $c(d)$ of the decision pattern d is defined by $c(d) = \sum_{k=1}^{K} c_k z_k(d)$.

The probability of decision pattern d can be expressed as

$$\Pr(d) = p(d_1) \cdots p(d_N) = \prod_{k=1}^{K} p_k^{z_k(d)},$$

where $p(d_n)$ is the probability of the alternative chosen by individual n, provided that the sample is an independent sample.

Cost-minimizing behavior is characterized by the following two axioms.

- AXIOM 1 *Independence of decision makers.* The decision maker chooses alternative k with probability p_k, independently of the choices made by the other decision makers.
- AXIOM 2 *Cost-Minimizing Behavior.* Samples with lower cost $c(d)$ are more probable.

Assume that we have two samples of the same size N. Let z_k^1 and z_k^2 denote the number of times alternative k is chosen in decision pattern (sample) 1 and 2, respectively. We arrive at the following definition of cost-minimizing behavior for the sample.

Definition 5 (Cost-Minimizing Behavior for Samples of N Independent Decision Makers). A probability distribution p represents cost-minimizing behavior, if and only if for any independent sample $d = (d_1, \ldots, d_N)$ of size N, the probability of observing d is a non increasing function of the total cost of the sample, hence $\Pr(d) = \prod_{k=1}^{K} p_k^{z_k(d)}$ is a non increasing function of $c(d) = \sum_{k=1}^{K} c_k z_k(d)$. In other words, $p(d)$ is cost minimizing if and only if

$$\sum_{k=1}^{K} c_k z_k^1 \leq \sum_{k=1}^{K} c_k z_k^2 \Longrightarrow \prod_{k=1}^{K} p_k^{z_k^1} \geq \prod_{k=1}^{K} p_k^{z_k^2}.$$

This is equivalent to Definition 1. According to Proposition 1 it determines completely the form of the probability distribution p as in (4.4). Hence, by defining cost-minimizing behavior for an N-sample of decision makers we obtain the choice probabilities for one single decision maker drawn at random from the population Ω. Thus AXIOM 1 together with AXIOM 2 determines the form of the probability distribution p completely. The reverse implication, that the logit model (4.4) implies AXIOM 2, is also true.

The total sum $c(d) = \sum_{n=1}^{N} c_{k_n}$ is a simple function. In addition it is easy to understand and it has great intuitive appeal. It is even difficult to imagine a cost-minimizing probability model which would not satisfy the definition above. If the members of the population Ω are cost minimizing when considered one at a time, then a sample of N members with lower total cost should be more probable. Note that the costs c_k could be replaced by a monotonic function of c_k without changing the content of the definition.

4.4.2 Axioms for Payoff-Maximizing Behavior

By replacing costs c_k with payoff values v_k and minimizing with maximizing AXIOM 2 is changed into:

- AXIOM 2a *Payoff-Maximizing Behavior.*
  Samples with higher total payoff $\sum_{k=1}^{K} z_k v_k$ are more probable.

Payoff maximizing behavior is characterized by the following definition:

Definition 6 (Payoff-Maximizing Behavior for Samples of N Independent Decision Makers). A probability distribution p represents payoff-maximizing behavior, if and only if for any independent sample $d = (d_1, \ldots, d_N)$ of size N, the probability of observing d is a non decreasing function of the total payoff of the sample, hence $\Pr(d) = \prod_{k=1}^{K} p_k^{z_k(d)}$ is a non decreasingfunction of

$v(d) = \sum_{k=1}^{K} v_k z_k(d)$. In other words, $p(d)$ is payoff maximizing if and only if

$$\sum_{k=1}^{K} v_k z_k^1 \le \sum_{k=1}^{K} v_k z_k^2 \Longrightarrow \prod_{k=1}^{K} p_k^{z_k^1} \le \prod_{k=1}^{K} p_k^{z_k^2}.$$

This is equivalent to Definition 2. According to Proposition 2 it determines completely the form of the probability distribution p, (4.7).

Hence, from Proposition 2, follows, that AXIOMS 1 and 2a imply that the probability distribution represents utility maximizing behavior in the sense of Definition 2. The reverse implication, that the logit model (4.7) implies that AXIOM 2a holds, is also true.

4.5 Axioms for ARUM Derivation

4.5.1 ARUM Derivation of the Simple (Multinomial) Logit Model

We shall now discuss the classical additive random perceived utility maximization derivation ARUM.

The classical approach to deriving the logit model postulates utility maximizing behavior. It is assumed that the behavior of a randomly chosen decision maker from the given population Ω of decision makers can be described by a probability distribution p over the choice set. The standard derivation assumes that the perceived utility of alternative k is composed of the sum of the payoff v_k and a random non observable component X_k. The decision maker is assumed to maximize his/her perceived utility and thus choosing the alternative that maximizes $(v_k + X_k)$. From the additional assumption that the random variables $X_1, \ldots, X_K$ are independently extreme value distributed follows the logit probabilities (4.7). This is the (additive) random utility maximizing approach, ARUM. It has been generalized in various directions. The classical approach can be axiomatized in the following way:

Let the perceived utility of alternative k for the decision maker be U_k, and let v_k denote the observable deterministic part of the utility – the payoff – and let X_k denote the unobservable random part.

The perceived utilities $[U_k]$ are assumed to be stochastic and linear in the observable and unobservable parts. The decision mechanism is deterministic: maximizing the random perceived utility.

- Axiom ARUM 1 *Independence of decision makers* The decision maker chooses alternative k with probability p_k, independently of the the other decision makers.
- Axiom ARUM 2 *Additivity* $U_k = v_k + X_k$, $k = 1, \ldots, K$.
- Axiom ARUM 3 *Independence of stochastic components* The random variables X_k, $k = 1, \ldots, K$ are independent.

- Axiom ARUM 4 *Extreme Value Distribution* The random variables X_k are distributed with the extreme value distribution $\Pr(X_k \leq x) = \exp(-\exp(-\theta x - E))$, $\theta > 0$, and where $E \approx 0.5772$ is Euler's constant.
- Axiom ARUM 5 *Maximizing Perceived Utility* The decision maker maximizes his/her perceived utility $U_k = v_k + X_k$.

Axiom ARUM 1 is the same as axiom AXIOM 1 above (Sect. 4.4.1), whereas axioms ARUM 2–5 replaces AXIOM 2a.

The axioms ARUM 1–5 imply that the probability distribution is the logit model (The proof is not given here. References are given in the notes, Sect. 4.8.):

$$p_k = \frac{\exp(\theta v_k)}{\sum_{k=1}^{K} \exp(\theta v_k)}. \tag{4.11}$$

Hence, from Proposition 2, follows, that Axiom ARUM 1–Axiom ARUM 5 imply that the probability distribution represents utility maximizing behavior in the sense of Definition 2. The reverse implication, that the logit model (4.11) implies that ARUM 2–5 hold, is false.

Payoff maximizing (utility) behavior PMB according to AXIOMS 1 and 2a – the approach used in this book although we usually formulate it in terms of cost-minimizing behavior – implies the logit model, (4.11). The same logit model is obtained if payoff (utility) maximizing behavior is defined by Axiom ARUM 1–5, the ARUM approach. Both approaches give the same logit model. However, there are two important differences between the two approaches. The first difference is that in the first approach the choice probability function is given by the logit model (4.11) if and only if AXIOMS 1 and 2a hold (according to Proposition 2). This implies that as soon as we use the logit formula we make the implicit assumption that AXIOMS 1 and 2a hold. Nothing of the sort can be said in the second approach; the logit formula follows from Axiom ARUM 1–5, but these axioms do not follow from the logit formula. The relationship between the two approaches can be illustrated as follows for Payoff Maximizing Behavior:

$$\text{AXIOM 1–2a} \Longleftrightarrow \text{Logit model} \Longleftarrow \text{ARUM 1–5}$$

The same relationship holds for cost-minimizing behavior.

The second difference between the two approaches is the following. Payoff (utility) maximizing behavior is formulated differently. In the first approach (PMB) payoff-maximizing behavior obtains if samples with greater value of the total payoff are more probable, i.e., are more often observed. This is something that can be refuted by observations if it is not true. The second approach (ARUM) assumes an intrinsic decision process based on maximizing (ARUM 5) the perceived payoff (utility) which in turn is prescribed to follow the details in Axiom ARUM 2–4. This is a detailed decision process the particularities of which cannot be refuted by observations, since the perceived payoffs (utilities) cannot be observed.

We have seen that the ARUM hypothesis implies the logit formula which in turn implies cost-minimizing behavior CMB or payoff-maximizing behavior PMB. The

ARUM formulation is therefore one example of specifying the detailed decision process which is otherwise left open when cost-minimizing CMB or payoff-maximization behavior PMB is postulated according to Definitions 1 and 2, respectively.

4.5.2 *Properties of the Expected Achieved Perceived Utility*

In this section we shall discuss the composite utility in the ARUM approach. The composite cost in our approach will be treated in Sect. 6.5.1.

The random variables X_k have expected value $E(X_k) = 0$, and variance $Var(X_k) = \pi^2/6\theta^2$. The expected achieved perceived utility $\tilde{v}(\theta)$, the so called composite utility, can be obtained:

$$E(maxU_k) = \tilde{v}(\theta) = \frac{1}{\theta} \log \sum_{k=1}^{K} \exp(\theta v_k). \tag{4.12}$$

If the extreme value distribution is written in the form

$$\Pr(X_k \leq x) = \exp(-\exp(-\theta x)),$$

then the expected achieved perceived utility is written $(1/\theta) \log \sum_{k=1}^{K} \exp(\theta v_k) + E/\theta$, where E is Euler's constant.

The expected achieved perceived utility is always greater than or equal to the expected payoff, $\tilde{v}(\theta) \geq \sum_{k=1}^{K} v_k p_k$.

Standard practice has been to use the expected achieved perceived utility $\tilde{v}(\theta)$, the composite utility, for evaluation purposes. This means adding to the expected payoff an arbitrary positive term. The practice can be motivated by including freedom of choice into a welfare measure (6.5), but this is usually not done.

As the parameter θ becomes small the expected achieved utility $\tilde{v}(\theta)$ grows without bound at the same time as the choice probabilities (4.11) become more equal. This reflects the fact that for small values of θ the payoff values v_k contribute little to the results. The decision making is instead governed by the extreme value distributed random variables X_k, whose variance tends to infinity as θ approaches zero. We shall see later that this behavior of the composite utility can be explained in terms of freedom of choice (Sect. 6.5).

As the parameter θ becomes large, on the other hand, the contribution to the decision making of the largest payoff value, $v_{k_{max}} = \arg[max_k v_k]$, is the only one that matters, and the expected achieved utility becomes equal to the expected utility, $\tilde{v}(\theta) = \sum_{k=1}^{K} v_k p_k = v_{k_{max}}$. This is a natural consequence of the fact that the variance of the extreme valued random variables X_k becomes zero in the limit as $\theta \to \infty$.

The discussion above shows that the influence on the decision making of the extreme value distributed random variables X_k goes from dominating completely to no importance at all when the parameter θ goes from zero to ∞. Thus there is an interval in which the value of θ influences the decision making. However, we shall see in Sect. 9.4 that the estimation by maximum likelihood of the parameter θ is based solely on the choice probabilities p_k, and does not refer to the extreme value distribution at all.

4.5.3 Generalized Extreme Value Model

Axiom ARUM 4 can be relaxed by replacing the extreme value distribution in Axiom ARUM 4 with a GEV distribution.

As before, let the perceived utility of alternative k for one decision maker be $U_k = v_k + X_k$, $k = 1, \ldots, K$. Let the random elements X_k have a simultaneous distribution function of the form

$$\Pr(X_1 \leq x_1, \ldots, X_K \leq x_K) = \exp(-G(\exp(-\theta x_1), \ldots, \exp(-\theta x_K))), \tag{4.13}$$

where the generator function G is any positive linearly homogeneous function which generates a cumulative distribution function and where $\theta > 0$. Hence, the random variables $[X_k]$ may be dependent. Again, let us assume that our decision maker maximizes his/her perceived utility. The probabilities p_k of alternative k being chosen can then be obtained:

Proposition 4 (Generalized Extreme Value Model). *For $k = 1, \ldots, K$, let the utility be $U_k = v_k + X_k$, $k = 1, \ldots, K$ and let the random component X_k of the utilities have the cumulative distribution function (4.13) with parameter $\theta > 0$, and assume that the decision maker maximizes his/her utility. Then the probability of alternative k being chosen is given by the GEV-model*

$$p_k = \frac{\exp(\theta v_k) G^k(\exp(\theta v_1), \ldots, \exp(\theta v_K))}{G(\exp(\theta v_1), \ldots, \exp(\theta v_K))}, \tag{4.14}$$

where G^k is the k-th partial derivative of G.

The proof is not given here. (For references see the notes, Sect. 4.8).

All GEV models are cost minimizing/payoff maximizing. We give this as a proposition:

Proposition 5 (GEV-Models are Cost Minimizing). *All GEV models defined by (4.13) are cost minimizing (payoff maximizing) in the sense of Definition 1 (Definition 2).*

Proof. We give the proof for the payoff (utility) maximizing case.

We shall show that we can find payoffs (utilities) such that the probabilities are given by (4.14).

Let the payoff (utility) be $(\theta v_k + \log G^k)$. Formula (4.6) in Definition 2 of maximizing payoff (utility) behavior becomes

$$\sum_{k=1}^{K}(\theta v_k + \log G^k)z_k^1 \geq \sum_{k=1}^{K}(\theta v_k + \log G^k)z_k^2 \Longrightarrow \prod_{k=1}^{K} p_k^{z_k^1} \geq \prod_{k=1}^{K} p_k^{z_k^2},$$

and it follows that

$$p_k = \frac{\exp(\gamma(\theta v_k + \log G^k))}{\sum_{k=1}^{K}\exp(\gamma(\theta v_k + \log G^k))} = \frac{\exp(\gamma\theta v_k)(G^k)^\gamma}{\sum_{k=1}^{K}\exp(\gamma\theta v_k)(G^k)^\gamma},$$

for some γ.

Putting $\gamma = 1$ we obtain (4.14) since the denominator $\sum_{k=1}^{K}\exp(\theta v_k)G^k$ is equal to $G(\exp(\theta v_1), \ldots, \exp(\theta v_K))$. □

The generator function G can be specified in various ways to give different, dependent or independent, models. We give two examples.

Let $G(\mathbf{x}) = \sum_{k=1}^{K} x_k$. Then $G^k = 1$ and we obtain once more the (multinomial) logit model (4.11),

$$p_k = \frac{\exp(\theta v_k)}{\sum_{k=1}^{K}\exp(\theta v_k)}, \quad k = 1, \ldots, K.$$

Again the probability represents payoff (utility) maximizing behavior (Proposition 2).

As another example, take the structured logit model with one cost measure, (5.11), which can be written as (5.12), i.e.,

$$p_{km} = \frac{\exp(\alpha_m - \gamma\tilde{c}_m(\gamma))}{\sum_m \exp(\alpha_m - \gamma\tilde{c}_m(\gamma))} \times \frac{\exp(-\gamma c_{km})}{\sum_k \exp(-\gamma c_{km})}. \tag{4.15}$$

This is a GEV-model that can be generated by $G(x) = \sum_{k=1}^{K}\exp(-\alpha_m)\sum_{m=1}^{M} x_{km}$, replacing the payoffs v_k in (4.14) with costs $-c_{km}$. This probability function represents cost-minimizing behavior according to Definition 12, Sect. 5.7.

4.6 Extensions

4.6.1 Comments on the Cost Function

Axiom 2 in Sect. 4.4 is formulated in terms of total cost $c(d)$. Is this the only function that can be used or is there another function $f(d)$? In statistical estimation theory a sufficient statistic is a function of the observations containing all information necessary for estimating a parameter. In our case the question is: is there a function $f(d)$ of the sample, which is sufficient in the sense that it can replace total cost $c(d)$ in Axiom 2? Indeed, under certain conditions, total cost $c(d)$ is the only function which is relevant in our case, as the lemma below shows.

Let $f(x) : \mathrm{R}^N \to \mathrm{R}$, be a differentiable function of $x = (x_1, \ldots, x_N)$ There are some natural requirements on f in our case.

1. The order of the costs in the sample should be of no importance. Hence the function f should be invariant under permutations of the elements of x.

2. A change of scale should change the value of f in proportion; f should be positive and linearly homogeneous; $f(\mu x) = \mu f(x),\ \mu > 0$.

3. The rate of change should be constant independently of n and the size of x_n;

$$\frac{\partial f}{\partial x_n} = \gamma, \quad n = 1, \ldots, N.$$

We have the following result:

Lemma 1. *Let $f(x) : \mathrm{R}^N \to \mathrm{R}$, be a differentiable function of $x = (x_1, \ldots, x_N)$ satisfying 1–3. Then $f = \gamma \sum_{n=1}^{N} x_n$ for some $\gamma > 0$.*

Proof. By Euler's formula we have

$$f = \sum_{n=1}^{N} x_n \frac{\partial f}{\partial x_n} = \gamma \sum_{n=1}^{N} x_n.$$

□

By the substitution $x_n = c_{k_n}$ we find that

$$f(c_{k_1}, \ldots, c_{k_N}) = \gamma \sum_{n=1}^{N} c_{k_n} = \gamma \sum_{k=1}^{K} z_k(d) c_k = \gamma c(d).$$

Hence, the cost $c(d)$, or a constant times this function, is the only function satisfying 1–3. With this in mind it is natural to define cost-minimizing behavior in terms of the cost function $c(d)$.

4.6.2 Different Interpretations of the Same Model

The definitions of cost-minimizing behavior and payoff (utility) maximizing behavior used in this book (Definitions 1 and 2) are broader than the ARUM model in the sense that very little has to be assumed about specific details of behavior in order to derive the probability distributions. Many different decision mechanisms are in principle compatible with the resulting probability distribution. We give two nested logit models as examples. Consider the structured logit model

$$p_{km} = \frac{\exp(-\gamma c_{km})}{\sum_{k=1}^{K} \sum_{m=1}^{M} \exp(-\gamma c_{km})}, \tag{4.16}$$

which is model (5.11) in Sect. 5.7 for $\alpha_m = 0$ (i.e., there are no activity constraints and only one cost condition). This can be rewritten in nested form as

$$p_{km} = \frac{\exp(-\tilde{c}_m(\gamma))}{\sum_{m=1}^{M} \exp(-\tilde{c}_m(\gamma))} \times \frac{\exp(-\gamma c_{km})}{\sum_{k=1}^{K} \exp(-\gamma c_{km})}.$$

where $\tilde{c}_m(\gamma) = -\frac{1}{\gamma} \log \sum_{k=1}^{K} \exp(-\gamma c_{km})$. This way of writing the model indicates that it can be interpreted as a two step model where alternative m is chosen first with probability

$$p_m = \frac{\exp(-\gamma \tilde{c}_m(\gamma))}{\sum_{m=1}^{M} \exp(-\gamma \tilde{c}_m(\gamma))}.$$

Then, conditionally on this choice, alternative k is chosen with probability

$$p_{k|m} = \frac{\exp(-\gamma c_{km})}{\sum_{k=1}^{K} \exp(-\gamma c_{km})}.$$

Hence the decisions are taken in two steps; first alternative m is chosen, then, conditionally on alternative m, alternative k is chosen. The probability p_{km} is made up of two steps, $p_{km} = p_m \times p_{k|m}$. This is a different decision mechanism than choosing (km) simultaneously as in (4.16). We can also interchange the order of the two steps obtaining

$$p_{km} = p_k \times p_{m|k} = \frac{-\exp(\gamma \tilde{c}_k(\gamma))}{\sum_{k=1}^{K} \exp(-\gamma \tilde{c}_k(\gamma))} \times \frac{\exp(-\gamma c_{km})}{\sum_{m=1}^{M} \exp(-\gamma c_{km})},$$

where $\tilde{c}_k(\gamma) = -\frac{1}{\gamma} \log \sum_{m=1}^{M} \exp(-\gamma c_{km})$.

Note that the model (4.16) has here been given three different interpretations. This can be empirically investigated by the methods given in Chap. 7.

We shall return to a discussion of the stepwise interpretation of structured (nested) logit models in Sects. 5.7.1 and 5.10.

4.6.3 Cost-Minimizing Behavior for one Decision Maker Making N Repeated Decisions

So far we have treated cost-minimizing behavior for a group/sample of independent decision makers. However, similar results can be obtained in the same way for *one* decision maker making a series of independent decisions. We shall use the same notation as before.

Consider one decision maker confronted with the choice between K alternatives denoted by $k,\ k = 1, \ldots, K$. Assume that the decision maker makes a series of N repeated independent decisions.

Let d_n be the decision taken by our decision maker at decision number n; d_n takes values in the choice set $[1, \ldots, K]$. Hence, $d_n = k$ if the decision maker chooses alternative k in the choice set at decision number n. Note that $n = 1, \ldots, N$ does not necessarily represent N consecutive time elements. The essential thing is that the sequence represents N independent decisions. Let the *decision pattern* $d = (d_1, \ldots, d_N)$ denote the decisions taken at the N repeated occasions.

We shall introduce a probability distribution $p = (p_1, \ldots, p_K)$ over the K alternatives in the choice set by defining

$$p_k = \Pr(\text{alternative } k \text{ is chosen in decision } n) = \Pr(d_n = k),\ k = 1, \ldots, K.$$

The probability distribution p takes care of variations in the decisions due to randomness or to factors unknown to us. These factors may very well be known to the decision maker, and hence his decision may be deterministic, but we choose to model them by including them in the random variation ascribed to the probability distribution p.

The probability of choosing alternative k is the same for each $n,\ n = 1, \ldots, N$. By using the definition and assuming independence between the decisions we can now write the probability of choosing the decision pattern $d = (d_1, \ldots, d_N)$ in the following way:

$$\Pr(d) = \Pr(d_1 = k_1, \ldots, d_N = k_N) = p_{k_1} \cdots p_{k_N}.$$

As before, let $z_k = z_k(d) =$ the number of times alternative k occurs in decision pattern d. Since the decisions are independent, the probability of the decision pattern d can be written

$$\Pr(d) = p_{k_1} \cdots p_{k_N} = \prod_{k=1}^{K} p_k^{z_k}.$$

Let now $c_k, k = 1, \ldots, K$, denote the cost of alternative k, and define the total cost of decision pattern d by $c(d) = \sum_{k=1}^{K} c_k z_k$. In the same way as before for a sample of independent decision makers we can define cost-minimizing behavior for repeated decisions by one decision maker. A probability distribution p represents cost-minimizing behavior, if and only if for any independent sample of independent repeated decisions for one decision maker, $d = (d_1, \ldots, d_N)$ of size N, the probability of observing d is a non increasing function of the total cost of the sample, hence $\Pr(d) = \prod_{k=1}^{K} p_k^{z_k(d)}$ is a non increasing function of $c(d) = \sum_{k=1}^{K} z_k(d) c_k$.

Consider two independent decision patterns d^1 and d^2 of the same size N. By letting $z^1 = (z_1^1, \ldots, z_K^1)$ and $z^2 = (z_1^2, \ldots, z_K^2)$ be the number of occurrences of alternatives $(1,\ldots,K)$ in decision patterns d^1 and d^2, respectively, we obtain in the same way as in Definition 1,

Definition 7 (Cost-Minimizing Behavior for One Decision Maker at N Repeated Decisions). A probability distribution $p=(p_1,\ldots,p_K)$ represents cost-minimizing behavior for repeated decisions by *one* decision maker if and only if, for any number N of repeated decisions,

$$c(d^1)\leq c(d^2)\Longrightarrow p(d^1)\geq p(d^2),$$

or equivalently

$$\sum_{k=1}^{K} c_k z_k^1 \leq \sum_{k=1}^{K} c_k z_k^2 \Longrightarrow \prod_{k=1}^{K} p_k^{z_k^1} \geq \prod_{k=1}^{K} p_k^{z_k^2}. \tag{4.17}$$

We shall say that a probability distribution p is a cost-minimizing probability distribution if it satisfies condition (4.17).

Given that we assume cost-minimizing behavior to hold, the form of the probability distribution can be derived exactly in the same way as in Proposition 1:

Proposition 6 (Cost-Minimizing Behavior for One Decision Maker at N Repeated Decisions). *The probability distribution $p=(p_1,\ldots,p_K)$ represents cost-minimizing behavior for one decision maker in the sense of Definition 7 if and only if it is a log linear probability distribution which can be written*

$$p_k=\frac{\exp(-\gamma c_k)}{\sum_{k=1}^{K}\exp(-\gamma c_k)}. \tag{4.18}$$

The choice probability distribution p for repeated decisions by *one* decision maker, (4.18), has exactly the same form as was given before for a random decision maker from a population of decision makers, (4.4). However, the costs c_k have a different interpretation. Here the costs c_k represents the same cost of alternative k at each decision occasion. Thus, the cost of alternative k is the same at each decision occasion and may be specific for the decision maker in question. Before, (4.4) in Proposition 1, the cost c_k designated the cost for the specific alternative k. This cost is the same for each decision maker in the population.

4.7 Comments

Ben-Akiva (1974), Williams (1977) and McFadden (1974, 1978a, 1981) introduced the now standard way of deriving the choice probabilities by assuming that the perceived utility of an alternative contains an additive unobservable random component with an extreme value distribution, i.e., the ARUM approach. From this follows logit type choice probabilities.

The use of the extreme value distribution (Weibull distribution, Weibull 1939, Gumbel distribution, Gumbel 1958) in the derivation of the logit model in the ARUM approach is usually motivated by its analytic convenience (Ben-Akiva and Lerman 1985, p. 104). A. Daly (1982) argues that "the extreme value distribution is the natural form of distribution to use, given that the alternatives actually modeled are themselves the result of choice over a large number of sub-alternatives (e.g. route, time of day). The utility of each alternative is then the maximum of the utilities of the sub-alternatives. Such a maximum can be shown to follow the distribution." However, Bell and Iida (1997) are of another opinion: "No convincing justification has been advanced for the assumption of Gumbel distributed utilities, other than that it yields a particularly tractable model, namely the logit model" (Bell and Iida 1997, p. 122). The approach used in this book avoids using extreme value distributed non observable random variables by showing how assumptions of cost-minimizing behavior according to Definition 1, leads to the classical models.

The assumption of a Gumbel distribution cannot be verified, since those random variables cannot be observed. The derivation used in this book of the logit model and other standard logit type models avoids this difficulty and permits testing the basic assumptions against observations.

In microeconomics utility functions play a dominant role in describing preferences (See e.g. McFadden 1981; Varian 1992). The non observable utilities introduced in the ARUM approach make it possible to formulate rational behavior by way of maximizing perceived utility. By using an extreme value distribution the expected achieved perceived utility can be calculated, $E(maxU_k) = \tilde{v}(\theta) = \frac{1}{\theta}\log\sum_{k=1}^{K}\exp(\theta v_k)$. This is an interesting and often used property of the ARUM approach. The expected achieved perceived utility $\tilde{v}(\theta)$ is always larger than the expected utility $v = \sum_{k=1}^{K} v_k p_k, \tilde{v}(\theta) \geq v = \sum_{k=1}^{K} v_k p_k$. However, here is a problem: what is the interpretation of the expected achieved perceived utility in the decision making? Perhaps the problem stands out more clearly if we instead talk about cost. We have expected achieved perceived cost $\tilde{c}(\beta) = -\frac{1}{\beta}\log\sum_{k=1}^{K}\exp(-\beta c_k)$ and at the same time expected cost $c = \sum_{k=1}^{K} c_k p_k$. Which one should be used in evaluating the decision process? If the model is correct, then the expected cost c tells which average costs we are going to observe. The composite cost, on the other hand, describes what we may expect when it comes to the achieved perceived cost. Which one is the more important? Is the average cost that we are going to observe more important or is the achieved perceived cost that come out of the derivation more important? In fact, we have the unusual situation that the perceived costs cannot be observed, but the expected achieved perceived cost can be calculated. This sounds as if we were in Alice's wonderland. The reason for this strange situation is the introduction of invisible costs which are added to the observable costs in an ad hoc manner.

The ARUM model fits nicely into microeconomic theory assuming that decision makers are rational utility maximizers. However, a serious problem with the

ARUM approach is the question of its identifiability from the observed distributions of demands since the extreme value distributed non observable components X_k cannot be observed. McFadden (1981) touches upon this question: "A related question is whether the model of individual utility maximization is identifiable from the observed distributions of demands, or whether other simpler or less restrictive models could generate the same observations." The notion of cost-minimizing behavior as used in this book avoids this problem. We shall see in Chap. 7 how the agreement between the assumptions of the model and empirical observations can be tested.

The definition of cost-minimizing behavior (Definition 1 in Sect. 4.2 and AXIOM 2 in Sect. 4.4.1) is weak but it cannot be substantially weakened , e.g., by assuming only that the choice probability function $\Pr(d)$ is decreasing in c_k.

The Chicago-man model, as the standard micro economic model is called by McFadden (1999), has serious difficulties when it comes to psychological evidence regarding human decision making (see also Gärling 1998; Rabin 1998). McFadden summarizes the properties of the Chicago-man model: it is convenient, it is successful, it is unnecessarily strong and it is false: "Almost all human behavior has a substantial rational component, at least in the broad sense of rationality. However, there is overwhelming behavioral evidence against a literal interpretation of Chicago-man as a universal model of choice behavior." (McFadden 1999). However, it may still be useful. "In an assessment of the role of psychological elements in travel demand analysis done for purposes of evaluating transportation policies, it is unclear whether one *needs* to incorporate these elements in order to obtain reliable predictions of behavior for policy purposes, or for that matter whether we are able of handling the resulting complexity when they are factored in. Economists and psychologists should recognize that what they consider the most interesting aspects of choice behavior are not necessarily important to transportation engineers. What is critical for transportation purposes is a "black box" that maps information about the transportation system into travel choices; the bottom line criterion is only that the black box works reliably. Every intervening construct within the box, such as an attitude, perception, or preference, is useful *only* if it is possible to provide both a mapping from information and experience to this construct *and* a mapping from this construct to choice that in tandem are more reliable than a direct "reduced form" mapping from experience and information to choice." (McFadden 2001).

The adherence to a black box which has been seen to produce useful results in the past may lead into difficulties when conditions change. The substitute of a black box for a model where the components can be verified and tested against observations can be misleading when analyzing future situations.

The approach used in this book, i.e., defining cost-minimizing behavior in terms of information about the choices made by a sample of decision makers is close to McFadden's ideas cited above. There is a direct mapping from information and experience of the behavior of the decision makers to choice, whereas in the classical ARUM derivation there is a theoretical construct in between, that cannot be observed, namely the extreme value distributed non observable components X_k. Also the additive structure presumed by adding observable and non observable components cannot be identified by observations. Both approaches result in the same form of the probability distributions for choice. Our approach is more direct and simpler.

However, the ARUM interpretation of the composite utility as the expected perceived achieved utility is in doubt, should we follow McFadden, since it relies on a theoretical construct that cannot be observed. We shall return to the behavior of the composite cost in Chap. 6.

4.8 Notes

Cost-minimizing behavior, Definition 1, goes back to the notion of Efficiency introduced for the gravity model by Smith (1978b) and further studied by Smith (1983, 1988) and Erlander and Smith (1990). Where in this book we talk about *cost-minimizing behavior*, Smith (1983) uses the macro *cost-efficiency principle*: "Under the conditions of the micro minimization hypothesis, one would expect on average to observe daily trip patterns with lower rather than higher levels of cumulative user-costs." " ... one may expect that pattern sequences with lower mean cost levels will be more likely. This is the essence of the macro *cost-efficiency principle* of travel behavior, " (Smith 1983). All results derived in this Chapter for cost minimizing behavior are obtained by using the corresponding results for the Efficiency approach. The notion of Activity Equivalence, here formulated as Definition 3, was introduced by Smith (1978b). Proposition 3 was given in a general setting by Smith (1983). The present proof of Proposition 3 was given by Erlander (1985) and in its most general form by Erlander and Smith (1990), see the Appendix, Sect. 9. See also Lundgren (1989), Erlander and Stewart (1990), Sen and Smith (1995) and Erlander and Lundgren (2004).

The first axiomatic derivation of the logit model was given by Luce (1959). The axioms for cost-minimizing behavior in Sect. 4.4, formulated in terms of efficient behavior, were given by Erlander (1998). The main contributions to the development of the classical ARUM approach (Sect. 4.5.1) are due to Ben-Akiva (1974), McFadden (1974, 1978a, 1981), Williams (1977), Williams and Senior (1977), and Daly and Zachary (1978). See also Ben-Akiva and Lerman (1985) and McFadden (2001). The correct definition of "inclusive value" or "composite cost" is due to Ben-Akiva (1974). The axioms of the ARUM model in Sect. 4.5.1 were discussed by Erlander (1998).

All discrete choice models that can be derived by the standard additive random utility maximizing approach (ARUM) with GEV distributions (McFadden 1974, 1978a, 1981) satisfy an appropriate variant of our definition of cost-minimizing behavior (Definition 1) or definition of payoff (utility) maximizing behavior (Definition 2). All such models can therefore be derived by our approach without having to assume the detailed structure of the decision making procedure exposed in Axioms 1–5 in Sect. 4.5.1. Hence our new definition offers a new interpretation of the underlying structure of these models.

The examples of GEV models at the end of Sect. 4.5.3 can be found e.g. in McFadden (1981).

Propositions 1 as well as 3 are similar to results by Farkas (Bachem and von Randow 1978). The vector $-\log p$ must lie in the polar cone spanned by the normals to the hyperplanes defined by the A and C matrices.

Chapter 5
Some Particular Logit Models

Abstract The new definition of cost-minimizing behavior is used to derive some discrete choice models – the stochastic route choice model, the multi-attribute discrete choice model, the gravity model for trip distribution, and the structured or nested logit model and others. The models can be extended and specialized in various directions.

5.1 Introduction

In this chapter we shall give examples of how the definition of cost-minimizing behavior can be used to derive some discrete choice models; amongst them the stochastic route choice model, the multi-attribute discrete choice model, the gravity model for trip distribution and the structured or nested logit model. The purpose is to give an idea of the richness and power of the new definition of cost-minimizing behavior. We have chosen a collection of discrete choice models commonly used in Transportation Science. The versions given here of the models are standard versions to illustrate the potential of the cost-minimizing approach.

We shall define cost-minimizing behavior for the different cases and obtain the particular form of the choice probability function by applying the general Representation Theorem, Proposition 3 in Sect. 4.3.

5.2 Stochastic Route Choice

We shall now discuss the simple stochastic route choice model with fixed link costs.. The case where link costs depend on link flow can also be treated by the approach used in this book, but this will be postponed until Chap. 8 in Part II.

Let there be N trip makers each one choosing one route for going from one location to another. Let there be K alternative routes and let the probability p_k, $k = 1, \ldots, K$ be the probability that route k is chosen. Let, furthermore, the (generalized) cost of using route k be c_k. In the absence of congestion and if the trip

S. Erlander, *Cost-Minimizing Choice Behavior in Transportation Planning*, Advances in Spatial Science, DOI 10.1007/978-3-642-11911-8_5,

makers have full information about costs we would expect all of them to choose the route with lowest cost. However, in reality not all trip makers choose the cheapest route, due to differences in information and differences in the evaluation of different routes. Still, we may assume that the behavior of the trip makers is cost minimizing, so that decision patterns with lower total sum of route costs will be observed more frequently, i.e. will be more probable. Assuming cost-minimizing behavior according to Definition 1 we find from Proposition 1 that the choice probabilities are given by

$$p_k = \frac{\exp(-\gamma c_k)}{\sum_{k=1}^{K} \exp(-\gamma c_k)}, \quad k = 1, \ldots, K.$$

This is the stochastic route choice model with fixed route costs.

5.3 The Multi-Attribute Discrete Choice Model

In the simple logit model treated in Sect. 4.2 there is one cost value, or generalized cost value, for each alternative in the choice set. The generalized cost may contain components such as monetary costs, time costs, parking costs etc. The generalized cost is obtained by adding the weighted components into one single cost measure. However, the components can also be treated as separate cost attributes, in which case the different weights will come out as coefficients. This is the Multi-Attribute Discrete Choice Model, that will be treated now. This model is also known under the name of Linear-in-Parameters Logit Model.

Assume that each alternative can be characterized by S attributes, and that we are given (deterministic) cost measures $c_{sk}, s = 1, \ldots, S$, for each alternative $k, k = 1, \ldots, K$. The cost measures c_{sk} can be of different types depending on the problem at hand: monetary costs such as outlays for gasoline, parking, transit fares; time "costs" such as walking time, in-vehicle travel time, and waiting time. Given a decision pattern $d = (d_1, \ldots, d_N)$ and the corresponding values $z_k, k = 1, \ldots, K$, of how many individuals that have chosen alternative k in this decision pattern, the total cost measure of attribute s for all N individuals in the sample of decision makers is given by $\sum_{k=1}^{K} z_k c_{sk}$. We can now express rationality of the decision makers in the sample by defining cost-minimizing behavior in terms of this total cost measure. We will state that a decision pattern for a sample that simultaneously has lower total cost measure for all S attributes would be more probable.

Let there be two independent samples of the same size N. We are going to compare the two decision patterns, d^1 and d^2, with corresponding values z_k^1 and z_k^2, $k = 1, \ldots, K$, for the samples. Cost-minimizing behavior is in this case defined as follows.

Definition 8 (Multi-Attribute Discrete Choice). A probability distribution $p = (p_1, \ldots, p_K)$ represents cost-minimizing behavior in the multi-attribute case if and only if, for any sample size N,

$$[\sum_{k=1}^{K} c_{sk} z_k^1 \leq \sum_{k=1}^{K} c_{sk} z_k^2, \quad s = 1, \ldots, S] \implies \prod_{k=1}^{K} p_k^{z_k^1} \geq \prod_{k=1}^{K} p_k^{z_k^2}. \tag{5.1}$$

The implication in (5.1) says that *if* the total cost measure for each attribute is lower in one sample compared to the other, the probability of observing the decisions of that sample would be higher. Note, as was observed already when discussing the general Representation Theorem for cost-minimizing probability distributions at the end of Sect. 4.3, that nothing is said about the relation between the probabilities of observing the decisions in the two samples, if the relations between the total cost measures do *not* hold. As before, to derive the form of the choice probability distribution it is enough to make assumptions on a restricted set of decision patterns only; namely the set of decision patterns satisfying (5.1). We can now give the conditions for cost-minimizing behavior in the multi-attribute case.

Proposition 7 (Multi-Attribute Discrete Choice Model). *The probability distribution* $p = (p_1, \ldots, p_K)$ *represents cost-minimizing behavior with respect to the cost measures* $c_{sk}, s = 1, \ldots, S, \ k = 1, \ldots, K$ *if and only if there exist* $\gamma_s \geq 0, \ \ s = 1, \ldots, S$*, such that*

$$p_k = \frac{\exp(-\sum_{s=1}^{S} \gamma_s c_{sk})}{\sum_{k=1}^{K} \exp(-\sum_{s=1}^{S} \gamma_s c_{sk})}, \quad k = 1, \ldots, K \tag{5.2}$$

Proof. By substitution into (5.1) we find that the multi-attribute logit model (5.2) satisfies the definition of a cost-minimizing probability distribution (Definition 8).

Assume that the choice probability distribution p is cost minimizing, i.e., (5.1) is satisfied. It then follows from the general Representation Theorem (Proposition 3 in Sect. 4.3) with the matrix A equal to the unity vector $[1, 1, \ldots, 1]^T$ and C equal to the matrix with elements $c_{sk}, \ s = 1, \ldots, S, \ k = 1, \ldots, K$, that the probability distribution p has the form (5.2). □

Hence, the multi-attribute discrete choice model (5.2) is not just cost minimizing according to Definition 8, but it is in fact the *only* probability distribution with this property with respect to S cost measures $[c_{sk}]$.

Note that the coefficients c_{sk} above represent cost attributes. If some attributes in fact denote payoff values the sign of the parameter should be reversed.

5.4 Generalized Cost

The expression $\sum_{s=1}^{S} \gamma_s c_{sk}$ in (5.2) can be looked upon as a generalized cost for alternative k. The total expected generalized cost is given by

$$c = \sum_{s=1}^{S} \sum_{k=1}^{K} \gamma_s c_{sk} \, p_k. \tag{5.3}$$

The term "generalized cost" stems from the fact that the influence by different factors can be expressed in terms of, say, the first cost by writing $c = \sum_{s=1}^{S} \gamma_s c_{sk} = \gamma_1 \sum_{s=1}^{S} (\gamma_s/\gamma_1) c_{sk} = \gamma_1 G_k$ where $G_k = \sum_{s=1}^{S} (\gamma_s/\gamma_1) c_{sk}$ and $\gamma_1 > 0$. Here G_k is the generalized cost measured in units of the first component. The expected generalized cost, expressed in terms of the first cost, is simply equal to the expected cost G defined by

$$G = \sum_{k=1}^{K} \sum_{s=1}^{S} (\gamma_s/\gamma_1) c_{sk} p_k.$$

If the coefficients are known the generalized cost is obtained directly. Otherwise the multi-attribute discrete choice model above offers a way of estimating the coefficients.

5.5 The Gravity Model for Trip Distribution

In our third example we consider the trip distribution problem.

In this case we have a sample of N trip makers (commuters) going from I origin zones to J destination zones. The trip makers correspond here to the decision makers in the previous examples. The decisions concern the simultaneous decisions about where to live and where to work, and the choice set consists of all pairs (i, j), $i = 1, \ldots, I$, and $j = 1, \ldots, J$. Let c_{ij} denote the (generalized) cost of going from zone i to zone j. Hence, c_{ij} corresponds to the generalized cost c_k in the previous examples. We want to derive the probability distribution $p = [p_{ij}]$, where p_{ij} is the probability that a trip maker chosen at random chooses origin i and destination j.

Assume that each of the N trip makers makes an independent decision, i.e. chooses one pair (i, j), independently of the other trip makers. Let $d_n = (i_n, j_n)$ be the decision taken by trip maker n, and let the *trip pattern* (decision pattern) $d = (d_1, \ldots, d_N)$ denote the decisions of all N trip makers. From this trip pattern we can compute the number of trip makers going from i to j, which we will denote T_{ij}. Hence, T_{ij} corresponds to z_k in the previous examples.

The probability of trip pattern d can now be expressed as

$$p(d) = \Pr(d_1 = (i_1, j_1), \ldots, d_N = (i_N, j_N)) = p_{i_1 j_1} \cdots p_{i_N j_N} = \prod_{i=1}^{I} \prod_{j=1}^{J} p_{ij}^{T_{ij}},$$

and the total cost of all the trips is given by

$$c(d) = c_{i_1 j_1} + \ldots + c_{i_N j_N} = \sum_{i=1}^{I} \sum_{j=1}^{J} c_{ij} T_{ij}.$$

To define cost-minimizing behavior for a sample we will consider two trip patterns d^1 and d^2 of the same size N, and the total number of trip makers going from i to j in the two trip patterns is given by the matrices $T^1 = [T_{ij}^1]$ and $T^2 = [T_{ij}^2]$.

Definition 9 (Activity Equivalence). We shall say that two trip patterns (samples) are *activity equivalent* if and only if

$$\sum_{j=1}^{J} T_{ij}^1 = \sum_{j=1}^{J} T_{ij}^2,\ i = 1, \ldots, I, \quad \text{and} \quad \sum_{i=1}^{I} T_{ij}^1 = \sum_{i=1}^{I} T_{ij}^2,\ j = 1, \ldots, J. \quad (5.4)$$

Activity equivalence means that the number of trip makers going to and from the zones $i = 1, \ldots, I$ and $j = 1, \ldots, J$, respectively, are equal for the two trip patterns. If not activity equivalent, it makes no sense comparing them with respect to cost. Now, if the trip makers are rational and tend to minimize cost, activity equivalent trip patterns (samples) with lower total cost would be more probable (frequently observed) than those with higher total cost. Note that in the definition of activity equivalence the constraints (5.4) have to be satisfied by the two trip patterns. The levels of the sums are not specified. In order to estimate the parameters of the probability in Proposition 8 the levels have to be specified as in (9.5) in the Appendix (Sect. 9.4).

Definition 10 (Cost-Minimizing Behavior in Trip Distribution). A probability distribution $p = [p_{ij}]$ represents cost-minimizing behavior if and only if for any two independent activity equivalent trip patterns, d^1 and d^2, of the same size N, we have

$$\sum_{i=1}^{I}\sum_{j=1}^{J} c_{ij} T_{ij}^1 \leq \sum_{i=1}^{I}\sum_{j=1}^{J} c_{ij} T_{ij}^2 \implies \prod_{i=1}^{I}\prod_{j=1}^{J} p_{ij}^{T_{ij}^1} \geq \prod_{i=1}^{I}\prod_{j=1}^{J} p_{ij}^{T_{ij}^2}. \quad (5.5)$$

It is again easy to see by substitution into (5.5) that the gravity model, $p_{ij} = \exp(\alpha_i + \beta_j - \gamma c_{ij})$, where $\gamma \geq 0$, satisfies the definition of cost-minimizing behavior, Definition 10.

Also, it follows again from the Representation Theorem (Proposition 3 in Sect. 4.3), that the probability distribution p represents a cost-minimizing probability distribution with respect to the marginal constraints (5.4) and the cost matrix $[c_{ij}]$ if and only if it has the gravity form above. The gravity model is the only probability distribution expressing cost-minimizing behavior (according to Definition 10) with respect to the trip costs $[c_{ij}]$ when the marginal trip numbers are identified as defining activity equivalence. We give this as a proposition.

Proposition 8 (Gravity Model for Trip Distribution). *The probability distribution $p = (p_{ij},\ i = 1, \ldots, I$ and $j = 1, \ldots, J,)$ represents cost-minimizing behavior with respect to the cost measures $(c_{ij},\ i = 1, \ldots, I$ and $j = 1, \ldots, J)$, if* *and only if there exist* $\alpha_i,\ i = 1, \dots, I\ \beta_j,\ j = 1, \dots, J$ *and* $\gamma \geq 0$ *such that*

$$p_{ij} = \exp(\alpha_i + \beta_j - \gamma c_{ij}),\ \ i = 1, \dots, I,\ j = 1, \dots, J \tag{5.6}$$

which can also be written

$$p_{ij} = \frac{\exp(\alpha_i + \beta_j - \gamma c_{ij})}{\sum_{i,j} \exp(\alpha_i + \beta_j - \gamma c_{ij})}. \tag{5.7}$$

Note that the parameters α_i and β_j will not be the same in (5.6) and (5.7), because a normalizing factor is included in the parameters in (5.6) whereas the factor is explicit in (5.7).

Proof. Substituting (5.6) into (5.4) and (5.5) we find that the gravity model (5.6) represents cost-minimizing behavior.

Assume now that the probability distribution p represents cost-minimizing behavior. We wish to derive the gravity model (5.6). The sample size N corresponds to the total number of trip makers T. The number z_k of decision makers taking decision k corresponds to T_{ij}. Let $z = (T_{11}, \dots, T_{1J}, T_{21}, \dots, T_{2J}, \dots T_{I1}, \dots, T_{IJ})^T$. In order to use the Representation Theorem for cost-minimizing probability distributions, Proposition 3, we need to identify the matrices A and C, and then show that the left hand side of (4.8) is satisfied.

The activity matrix A is here the transportation matrix

$$\mathrm{A} = \left\{ \begin{array}{cccccccccccccc}
1 & 1 & \cdots & 1 & 0 & 0 & \cdots & 0 & \cdots & 0 & 0 & \cdots & 0 \\
0 & 0 & \cdots & 0 & 1 & 1 & \cdots & 1 & \cdots & 0 & 0 & \cdots & 0 \\
 & \vdots & & & & & & \vdots & \ddots & & & \vdots & \\
0 & 0 & \cdots & 0 & 0 & 0 & \cdots & 0 & \cdots & 1 & 1 & \cdots & 1 \\
 & & & & & & & & & & & & \\
1 & 0 & \cdots & 0 & 1 & 0 & \cdots & 0 & \cdots & 1 & 0 & \cdots & 0 \\
0 & 1 & \cdots & 0 & 0 & 1 & \cdots & 0 & \cdots & 0 & 1 & \cdots & 0 \\
\vdots & & \ddots & \vdots & \vdots & & \ddots & \vdots & \cdots & \vdots & & \ddots & \vdots \\
0 & 0 & \cdots & 1 & 0 & 0 & \cdots & 1 & \cdots & 0 & 0 & \cdots & 1
\end{array} \right\}.$$

The cost matrix C is the single vector consisting of the concatenated elements (c_{ij}). Hence,

$$\mathrm{A}z^1 = \mathrm{A}z^2 \iff \left\{ \begin{array}{c}
\sum_{j=1}^{J} T_{1j}^1 = \sum_{j=1}^{J} T_{1j}^2 \\
\vdots \\
\sum_{j=1}^{J} T_{Ij}^1 = \sum_{j=1}^{J} T_{Ij}^2 \\
\sum_{i=1}^{I} T_{i1}^1 = \sum_{i=1}^{I} T_{i1}^2 \\
\vdots \\
\sum_{i=1}^{I} T_{iJ}^1 = \sum_{i=1}^{I} T_{iJ}^2
\end{array} \right\},$$

and

$$Cz^1 \leq Cz^2 \Longleftrightarrow \sum_{i=1}^{I}\sum_{j=1}^{J} c_{ij}T_{ij}^1 \leq \sum_{i=1}^{I}\sum_{j=1}^{J} c_{ij}T_{ij}^2.$$

Hence, the activity conditions are satisfied, and from Proposition 3 follows the gravity model (5.6). □

Note that the parameters α_i and β_j are not unique, since there is one linear relationship in the A matrix because of $\sum_{ij} T_{ij} = \sum_{ij} z_{ij} = N$. In practice it is usually enough to fix the value of one of them, say $\alpha = 1$, to obtain uniqueness. (The full conditions for uniqueness are given in Theorem 1 in the Appendix.)

5.6 The Gravity Model for Trip Distribution with Several Cost Attributes

If there are several cost attributes we can combine the multi-attribute and the gravity model into one model treating all cost attributes at a time. Let $c_{sij}, s = 1, \ldots, S,\ i = 1, \ldots, I,\ j = 1, \ldots, J$, denote the value of attribute s for trips going from i to j. Combining Definitions 8 and 10 we obtain the new definition below of cost minimizing behavior for the case with several cost attributes. Activity equivalence is defined as before (5.4).

Definition 11 (Cost-Minimizing Behavior in Trip Distribution with Several Cost Attributes). A probability distribution $p = (p_{ij})$ represents cost-minimizing behavior if and only if for any two independent activity equivalent trip patterns, d^1 and d^2, of the same size N, we have

$$\sum_{i=1}^{I}\sum_{j=1}^{J} c_{sij}T_{ij}^1 \leq \sum_{i=1}^{I}\sum_{j=1}^{J} c_{sij}T_{ij}^2,\ s = 1, \ldots, S \Longrightarrow \prod_{i=1}^{I}\prod_{j=1}^{J} p_{ij}^{T_{ij}^1} \geq \prod_{i=1}^{I}\prod_{j=1}^{J} p_{ij}^{T_{ij}^2}. \tag{5.8}$$

The implication in (5.8) says that *if* the total cost measure for each attribute is lower in one sample compared to the other, the probability of observing the decisions of that sample should be higher. Note that nothing is said about the relation between the probabilities of observing the decisions in the two samples, should the relations between the total cost measures *not* hold.

It is again easy to see by substitution into (5.8) that the gravity model,

$$p_{ij} = \exp(\alpha_i + \beta_j - \sum_{s=1}^{S} \gamma_s c_{sij}), \tag{5.9}$$

where $\gamma_s \geq 0,\ s = 1, \ldots, S$, satisfies the definition of cost-minimizing probability distribution, Definition 11. This is the gravity model with several cost attributes.

By identifying the matrices A and C

$$\mathrm{A}z^1 = \mathrm{A}z^2 \Longleftrightarrow \left\{ \begin{array}{c} \sum_{j=1}^{J} T_{1j}^1 = \sum_{j=1}^{J} T_{1j}^2 \\ \vdots \\ \sum_{j=1}^{J} T_{Ij}^1 = \sum_{j=1}^{J} T_{Ij}^2 \\ \sum_{i=1}^{I} T_{i1}^1 = \sum_{i=1}^{I} T_{i1}^2 \\ \vdots \\ \sum_{i=1}^{I} T_{iJ}^1 = \sum_{i=1}^{I} T_{iJ}^2 \end{array} \right\},$$

and

$$\mathrm{C}z^1 \le \mathrm{C}z^2 \Longleftrightarrow \left\{ \begin{array}{c} \sum_{i=1}^{I} \sum_{j=1}^{J} c_{1ij} T_{ij}^1 \le \sum_{i=1}^{I} \sum_{j=1}^{J} c_{1ij} T_{ij}^2 \\ \vdots \\ \sum_{i=1}^{I} \sum_{j=1}^{J} c_{Sij} T_{ij}^1 \le \sum_{i=1}^{I} \sum_{j=1}^{J} c_{Sij} T_{ij}^2 \end{array} \right\},$$

it follows again from the Representation Theorem (Proposition 3 in Sect. 4.3), that the probability distribution p represents a cost-minimizing probability distribution with respect to the marginal constraints (5.4) and the cost matrices $[c_{sij}]$ if and only if it has the gravity form (5.9). The gravity model is the only probability distribution expressing cost-minimizing behavior (according to Definition 11) with respect to the trip costs $[c_{sij}]$ when the marginal trip numbers are identified as defining activity equivalence. We give this as a proposition.

Proposition 9 (Gravity Model for Trip Distribution with Several Cost Attributes). *The probability distribution* $p = (p_{ij}, i = 1, \ldots, I$ *and* $j = 1, \ldots, J)$ *represents cost-minimizing behavior with respect to the cost measures* c_{sij}, $s = 1, \ldots, S$, $i = 1, \ldots, I$ *and* $j = 1, \ldots, J$, *if and only if there exist* α_i, $i = 1, \ldots, I$, β_j, $j = 1, \ldots, J$ *and* $\gamma_s \ge 0$, $s = 1, \ldots, S$ *such that*

$$p_{ij} = \exp(\alpha_i + \beta_j - \sum_{s=1}^{S} \gamma_s c_{sij}), \quad i = 1, \ldots, I, \; j = 1, \ldots, J$$

which can also be written

$$p_{ij} = \frac{\exp(\alpha_i + \beta_j - \sum_{s=1}^{S} \gamma_s c_{sij})}{\sum_{i,j}^{K} \exp(\alpha_i + \beta_j - \sum_{s=1}^{S} \gamma_s c_{sij})}.$$

Proof. The proof parallels the proof of Proposition 8. □

The sum $\sum_{s=1}^{S} \gamma_s c_{sij}$ can be replaced with a generalized cost according to Sect. 5.4 if the coefficients are known.

The two propositions about the gravity model for trip distribution illustrates how general models can be constructed. Activity constraints are formulated as equalities leading to parameters/multipliers undetermined in sign. Cost conditions are formulated as inequalities resulting in non negative parameters/multipliers. Note that there is no essential difference in the mathematical treatment of equalities defining activity constraints and inequalities defining cost conditions.

5.7 Structured (Nested) Logit Models

In the simple logit model (4.4) all alternatives are treated equally. The only thing that matters is the (generalized) cost of the alternative. There are many situations where a hierarchical structure is appropriate. For example in modeling choice of travel mode and destination it may be argued that there are hidden differences between modes not covered by costs alone. In this case we would like to compare costs within each mode. This can be done by using structured models. There are many ways of formulating structured models. We shall give one simple structured logit model in two versions and also the standard nested logit model in two versions. The latter is also derived by the standard ARUM approach using generalized extreme value distributions.

Proposition 3 showed that there is a one-to-one correspondence between cost minimizing behavior and log linear probability functions. Hence all probability functions of the form

$$p_k = \exp(\mu + \alpha^T \mathbf{a}_k - \gamma^T \mathbf{c}_k),\ k = 1, \ldots, K,$$

represent cost-minimizing behavior with respect to the activity matrix A and the cost matrix C. Common to all such log linear models is that they can be reformulated as structured (nested) logit models by using composite cost functions. One example was given for the simple logit model in Sect. 4.6.2. In the following we shall utilize this approach.

5.7.0.1 Blue/Red Bus Example

To introduce the ideas we shall start with a very simple model; the blue/red bus case. Let there be two bus lines A and B connecting two cities. Let the cost of the trip from one city to the other be equal to c on both lines. Let z_A and z_B be the number of trip makers taking line A and B respectively. Let p_A and p_B denote the choice probabilities. Take two samples of the same size N; (z_A^1, z_B^1) and (z_A^2, z_B^2). Define cost-minimizing behavior to obtain if

$$cz_A^1 + cz_B^1 \leq cz_A^2 + cz_B^2 \Longrightarrow p_A^{z_A^1} p_B^{z_B^1} \geq p_A^{z_A^2} p_B^{z_B^2},$$

i.e., if total cost of the first sample is less than total cost of the second sample then the likelihood of the first sample is higher. Cost-minimizing behavior implies the following choice probabilities:

$$p_A = p_B = \frac{\exp(-\beta c)}{\exp(-\beta c) + \exp(-\beta c)} = \frac{1}{2}.$$

Now, paint half of the B buses blue and half of them red. Then we have three bus lines, A, B and R. The corresponding logit model becomes

$$p_A = p_B = p_R = \frac{\exp(-\beta c)}{\exp(-\beta c) + \exp(-\beta c) + \exp(-\beta c)} = \frac{1}{3}.$$

The trip makers distribute themselves evenly over the three bus lines. However, lines B and R are identical except for the color. Thus we need to structure the logit model to take this into account. This can be done by introducing a constraint in the definition of cost-minimizing behavior. Cost-minimizing behavior is defined as follows;

$$\left.\begin{array}{l} z_B^1 + z_R^1 = z_B^2 + z_R^2 \text{ and} \\ \\ cz_A^1 + cz_B^1 + cz_R^1 \le cz_A^2 + cz_B^2 + cz_R^2, \end{array}\right\} \Longrightarrow p_A^{z_A^1} p_B^{z_B^1} p_B^{z_R^1} \ge p_A^{z_A^2} p_B^{z_B^2} p_B^{z_R^2}.$$

The constraint means that we are comparing only samples with the same number of trip makers taking bus lines B and R.

Now the logit model becomes

$$p_A = \frac{\exp(-\beta c)}{\exp(-\beta c) + \exp(\alpha - \beta c) + \exp(\alpha - \beta c)}$$

$$p_B = p_R = \frac{\exp(\alpha - \beta c)}{\exp(-\beta c) + \exp(\alpha - \beta c) + \exp(\alpha - \beta c)}.$$

From this we have

$$p_B = p_R = \exp(\alpha) p_A.$$

If $\alpha = 0$, then the trips are evenly distributed over the three lines, $p_A = p_B = p_R = 1/3$. If $\exp(\alpha) = 1/2$, then $p_B + p_R = p_A$, and we are back at the first situation where half of the trip makers choose A and half choose B+R; $p_A = 1/2$, $p_B = p_R = 1/4$. Thus we see that by introducing a constraint in the definition of cost-minimizing behavior we can handle cases where some alternatives are more similar than others. We can even handle the case where the alternatives are identical. In application situations the observed frequencies will indicate through estimates of the parameter α which model is appropriate. We shall now return to the general case.

5.7.1 The Structured Logit Model: The Joint Logit Model

Structured logit models are multilevel models where, in the simple two-level case treated here, there is a choice between M alternatives $m = 1, \ldots, M$ in the upper level, and, for each choice in the upper level, there is a choice between K alternatives $k = 1, \ldots, K$ in the lower level. The upper level choice may be the choice of mode of transportation and the lower level choice may be the choice of destination. The division into one upper level and one lower level does not indicate that the decisions in the upper and the lower levels respectively are taken in this order. We assume that the decisions about mode and destination are taken simultaneously. Structured models are sometimes given the interpretation that upper level decisions precede lower level decisions. This may be natural in some cases, but we do not presume this here. We shall comment on this later. We shall give two versions of the joint logit model.

The idea is now to introduce activity constraints in such a way that the costs are compared for similar trip patterns only. E.g. if the decision in the upper level deals with choosing between public transport and private car it is desirable to restrict comparisons to trip patterns having the same number of trip makers choosing public transport and the same number choosing private car. In this way we take to some extent into consideration the fact that there may be other differences between public transport and private car than those related to cost and other attributes. The consequence of the introduction of activity constraints in this way will, as we shall see, be that in the choice probability function there is a public transport parameter and a private car parameter in addition to the cost parameter.

Let there be costs c_{km}, $k = 1, \ldots, K$, and $m = 1, \ldots, M$, and let z_m and z_{km} be the number of times alternative m and km are chosen in the upper and lower levels, respectively. Clearly, $z_m = \sum_{k=1}^{K} z_{km}$. Assume that we have two samples of the same size $N = \sum_{m=1}^{M} z_m$. Let z_m^1, z_{km}^1 and z_m^2, z_{km}^2 denote the number of times alternatives m and km are chosen in decision sample 1 and 2, respectively.

In order to compare costs for trip patterns that are similar with respect to properties not covered by cost attributes we shall introduce activity equivalence constraints,

$$\sum_{k=1}^{K} z_{km}^1 = \sum_{k=1}^{K} z_{km}^2, \quad m = 1, \ldots, M.$$

This means that we restrict comparisons to samples with equal number of times alternative $m, m = 1, \ldots, M$, are chosen.

We then define cost-minimizing behavior in the following way:

Definition 12 (Cost-Minimizing Behavior in the Structured Logit Model with One Cost Constraint). A probability distribution $p = (p_{km}, k = 1, \ldots, K, m = 1, \ldots, M)$ represents cost-minimizing behavior if and only if, for any sample size N,

$$\left.\begin{array}{l}\sum_{k=1}^{K} z_{km}^{1} = \sum_{k=1}^{K} z_{km}^{2},\ m = 1,\ldots,M, \text{ and} \\ \sum_{k=1}^{K}\sum_{m=1}^{M} c_{km} z_{km}^{1} \leq \sum_{k=1}^{K}\sum_{m=1}^{M} c_{km} z_{km}^{2},\end{array}\right\}$$

$$\Longrightarrow \prod_{k=1}^{K}\prod_{m=1}^{M} p_{km}^{z_{km}^{1}} \geq \prod_{k=1}^{K}\prod_{m=1}^{M} p_{km}^{z_{km}^{2}}.$$

The content of the definition is that cost-minimizing behavior prevails if activity equivalent samples (i.e. samples with equal number of times that decision m is chosen for each $m = 1,\ldots,M$) with lower values of total cost are more likely and will be more frequently observed. The activity constraints will generate parameters α_m, $m = 1,\ldots,M$, and the cost inequality will generate a parameter γ. The parameters play a role similar to Lagrange multipliers; a high value indicates that the condition is active. This is a structured logit model. It takes care to some extent of (hidden) similarities within each upper level alternative m. In this sense the model is hierarchical. However, the model does not presume any specific order of the decision making.

We shall use our general results in Proposition 3 in the same way as when deriving the gravity model in Proposition 8. Let

$$z = (z_{11},\ldots,z_{1K},z_{21},\ldots,z_{2K},\ldots,z_{M1},\ldots,z_{MK})^{T}.$$

The cost matrix C is the single vector consisting of the concatenated elements (c_{km}), $\mathrm{C} = (c_{11},\ldots,c_{1K},c_{21},\ldots,c_{2K},\ldots,c_{M1},\ldots,c_{MK})$ and the activity matrix A is

$$\mathrm{A} = \left\{\begin{array}{cccc} 1\,1\cdots 1 & 0\,0\cdots 0 & \cdots & 0\,0\cdots 0 \\ 0\,0\cdots 0 & 1\,1\cdots 1 & \cdots & 0\,0\cdots 0 \\ \vdots & \vdots & \ddots & \vdots \\ 0\,0\cdots 0 & 0\,0\cdots 0 & \cdots & 1\,1\cdots 1 \end{array}\right\}.$$

The conditions in Definition 12 become

$$\mathrm{A}z^1 = \mathrm{A}z^2 \text{ and } \mathrm{C}z^1 \leq \mathrm{C}z^2 \Longrightarrow \prod_{k=1}^{K}\prod_{m=1}^{M} p_{km}^{z_{km}^{1}} \geq \prod_{k=1}^{K}\prod_{m=1}^{M} p_{km}^{z_{km}^{2}}.$$

We obtain the following proposition.

Proposition 10 (Cost-Minimizing Behavior in the Structured Logit Model with One Cost Constraint). *The probability distribution* $p = (p_{km}, k = 1,\ldots,K, m = 1,\ldots,M)$ *represents cost-minimizing behavior with respect to the cost measures* $(c_{km}, k = 1,\ldots,K$ *and* $m = 1,\ldots,M)$, *if and only if there exist* μ, $\alpha_m, m = 1,\ldots,M$ *and* $\gamma \geq 0$ *such that*

$$p_{km} = \exp(\mu + \alpha_m - \gamma c_{km}),\ \ k = 1,\ldots,K,\ m = 1,\ldots,M, \qquad (5.10)$$

which can also be written

$$p_{km} = \frac{\exp(\alpha_m - \gamma c_{km})}{\sum_{k=1}^{K}\sum_{m=1}^{M}\exp(\alpha_m - \gamma c_{km})} \quad k = 1,\ldots,K,\ m = 1,\ldots,M. \tag{5.11}$$

Proof. The proposition follows from our general results (Proposition 3). □

This is the joint logit model. Note that here the mode choice constants α_m can be viewed as Lagrange multipliers related to the activity constraints.

By using the composite cost,

$$\tilde{c}_m(\gamma) = -\frac{1}{\gamma}\log\sum_{k=1}^{K}\exp(-\gamma c_{km}),\ m = 1,\ldots,M$$

(5.11) can be written

$$p_{km} = \frac{\exp(\alpha_m - \gamma\tilde{c}_m(\gamma))}{\sum_{m=1}^{M}\exp(\alpha_m - \gamma\tilde{c}_m(\gamma))} \times \frac{\exp(-\gamma c_{km})}{\sum_{k=1}^{K}\exp(-\gamma c_{km})}. \tag{5.12}$$

This is one example of the general observation that logit models often can be given a "nested" form. We shall comment upon this way of writing the formula after presenting the next version of the joint logit model.

A second version of the structured logit model is obtained if in the definition we separate several measures of costs (or attributes), i.e. one constraint for each m, $m = 1,\ldots,M$.

Definition 13 (Cost-Minimizing Behavior in the Structured Logit Model with Several Cost Constraints). A probability distribution $p = (p_{km})$, $k = 1,\ldots,K$, $m = 1,\ldots,M$, represents cost-minimizing behavior if and only if for any two independent activity equivalent trip patterns, d^1 and d^2, of the same size N, we have

$$\left.\begin{aligned} &\textstyle\sum_{k=1}^{K} z_{km}^1 = \sum_{k=1}^{K} z_{km}^2,\ m = 1,\ldots,M, \text{ and} \\ &\textstyle\sum_{k=1}^{K} c_{km} z_{km}^1 \le \sum_{k=1}^{K} c_{km} z_{km}^2,\ m = 1,\ldots,M, \end{aligned}\right\}$$
$$\Longrightarrow \prod_{k=1}^{K}\prod_{m=1}^{M} p_{km}^{z_{km}^1} \ge \prod_{k=1}^{K}\prod_{m=1}^{M} p_{km}^{z_{km}^2}.$$

The content of the definition is that cost-minimizing behavior prevails if activity equivalent samples (i.e. samples with equal number of times that decision m is chosen for each $m = 1,\ldots,M$) with lower values of total cost for each $m = 1,\ldots,M$ are more likely and will be more frequently observed. The activity constraints will generate parameters α_m, $m = 1,\ldots,M$, and the cost inequalities will generate parameters γ_m, $m = 1,\ldots,M$. The parameters play a role similar to Lagrange multipliers; a high value indicates that the condition is active. Again, this is a structured logit model. It takes care to some extent of (hidden) similarities within each upper level alternative m. In this sense the model is hierarchical. However, the model does not presume any specific order of the decision making.

Proposition 11 (Cost-Minimizing Behavior in the Structured Logit Model with Several Cost Constraints). *The probability distribution $p = (p_{km}, k = 1, \ldots, K, m = 1, \ldots, M)$ represents cost-minimizing behavior with respect to the cost measures $(c_{km}, k = 1, \ldots, K$ and $m = 1, \ldots, M)$, if and only if there exist μ, α_m, $m = 1, \ldots, M$ and $\gamma_m \geq 0$ such that*

$$p_{km} = \exp(\mu + \alpha_m - \gamma_m c_{km}), \quad k = 1, \ldots, K, \; m = 1, \ldots, M,$$

which can also be written

$$p_{km} = \frac{\exp(\alpha_m - \gamma_m c_{km})}{\sum_{k=1}^{K} \sum_{m=1}^{M} \exp(\alpha_m - \gamma_m c_{km})} \quad k = 1, \ldots, K, \; m = 1, \ldots, M. \tag{5.13}$$

Introducing one cost constraint for each m changes the parameter γ into $\gamma_1, \ldots \gamma_M$.

Proof. The proposition follows from our general results (Proposition 3), by letting the cost matrix be the matrix

$$C = \left\{ \begin{matrix} c_{11} & c_{12} & \cdots & c_{1K} & 0 & 0 & \cdots & 0 & \cdots & 0 & 0 & \cdots & 0 \\ 0 & 0 & \cdots & 0 & c_{21} & c_{22} & \cdots & c_{2K} & \cdots & 0 & 0 & \cdots & 0 \\ & & \vdots & & & & \vdots & & \ddots & & & \vdots & \\ 0 & 0 & \cdots & 0 & 0 & 0 & \cdots & 0 & \cdots & c_{M1} & c_{M2} & \cdots & c_{MK} \end{matrix} \right\}.$$

□

By using the composite costs,

$$\tilde{c}_m(\gamma_m) = -\frac{1}{\gamma_m} \log \sum_{k=1}^{K} \exp(-\gamma_m c_{km}), \; m = 1, \ldots, M,$$

(5.13) can be written

$$p_{km} = \frac{\exp(\alpha_m - \gamma_m \tilde{c}_m(\gamma_m))}{\sum_{m=1}^{M} \exp(\alpha_m - \gamma_m \tilde{c}_m(\gamma_m))} \times \frac{\exp(-\gamma_m c_{km})}{\sum_{k=1}^{K} \exp(-\gamma_m c_{km})}. \tag{5.14}$$

This way of writing the choice probabilities indicates why it is often called a "nested logit model", since the choice probabilities p_{km} can be written as

$$p_{km} = p_m \times p_{k|m}, \tag{5.15}$$

where p_m is the marginal probability of choosing alternative m in the upper level, and $p_{k|m}$ is the conditional probability of choosing alternative k in the lower level provided that alternative m is chosen for the upper level. In the structured model we can identify the factors in (5.14) as

$$p_m = \frac{\exp(\alpha_m - \gamma_m \tilde{c}_m(\gamma_m))}{\sum_{m=1}^{M} \exp(\alpha_m - \gamma_m \tilde{c}_m(\gamma_m))},$$

and

$$p_{k|m} = \frac{\exp(-\gamma_m c_{km})}{\sum_{k=1}^{K} \exp(-\gamma_m c_{km})}.$$

The interpretation is often made that the choices on the upper m level precede the decisions on the lower level k. However, there is no assumption about the order of the decisions made in the derivation of the formula.

Writing the choice probabilities in the form (5.15) gives the impression that the decisions in the lower level are made with the only reference to the upper level that m is chosen in the upper level. In the upper level, on the other hand, decisions are influenced by what can be anticipated in the lower level. This is accomplished by using the composite cost. This means that the decisions in the upper and lower levels are coupled through the use of the composite cost. The decisions are not independent, in spite of the fact that (5.15) may suggest independence. This should be borne in mind when using (5.12) and (5.14). These formulas do not in general explain behavior better than the simpler formulas (5.11) and (5.13).

We have presented the joint logit model in a two-level version. Hierarchical models with more levels can be constructed by adding more constraints in the definition of activity equivalence.

See also Sect. 4.6.2.

5.7.2 The Standard Nested Logit Model

The standard nested logit model is an hierarchical model where composite cost is used to obtain coupling between consecutive levels in the same way as in 5.12 and 5.14. It is instructive to see what has to be changed in order to derive the standard nested logit model from cost-minimizing behavior. As can be anticipated the introduction of additional cost constraints is enough. Adding a cost constraint expressed in terms of composite costs $\tilde{c}_m(\beta) = -\frac{1}{\beta} \log \sum_{k=1}^{K} \exp(-\beta c_{km})$ in the definition of cost-minimizing behavior (Definition 12) results in the standard nested logit model. This can be done in several ways. We give the most straightforward formulation.

Definition 14 (Cost-Minimizing Behavior in the Standard Nested Logit Model with One Constraint on Composite Cost). A probability distribution $p = (p_{km}, k = 1, \ldots, K, m = 1, \ldots, M)$ represents cost-minimizing behavior if and only if, for any sample size N,

$$\left.\begin{array}{l}\sum_{k=1}^{K} z_{km}^{1} = \sum_{k=1}^{K} z_{km}^{2},\ m = 1,\ldots,M,\\ \sum_{k=1}^{K}\sum_{m=1}^{M} c_{km} z_{km}^{1} \le \sum_{k=1}^{K}\sum_{m=1}^{M} c_{km} z_{km}^{2},\\ \sum_{m=1}^{M} \tilde{c}_m(\beta) z_m^{1} \le \sum_{m=1}^{M} \tilde{c}_m(\beta) z_m^{2},\end{array}\right\}$$
$$\Longrightarrow \prod_{k=1}^{K}\prod_{m=1}^{M} p_{km}^{z_{km}^{1}} \ge \prod_{k=1}^{K}\prod_{m=1}^{M} p_{km}^{z_{km}^{2}}.$$

The addition of the third inequality means that the two levels are not treated in a symmetric way. The choice of an alternative m in the upper level is governed not only by the costs c_{km} but by the composite cost $\tilde{c}_m(\beta)$ as well. In addition to the parameters α_m, $m = 1,\ldots,M$, and β the third inequality generates an additional parameter γ.

Proposition 12 (Cost-Minimizing Behavior in the Standard Nested Logit Model with One Constraint on Composite Cost). *The probability distribution* $p = (p_{km},\ k = 1,\ldots,K,\ m = 1,\ldots,M)$ *represents cost-minimizing behavior with respect to the cost measures* $(c_{km},\ k = 1,\ldots,K$ *and* $m = 1,\ldots,M)$ *and the composite costs* $\tilde{c}_m(\gamma)$, $m = 1,\ldots,M$, *if and only if there exist* μ, α_m, $m = 1,\ldots,M$ *and* $\beta \ge 0$ *and* $\gamma \ge 0$ *such that*

$$p_{km} = \exp(\mu + \alpha_m - \beta c_{km} - \gamma \tilde{c}_m(\beta)),\ \ k = 1,\ldots,K,\ m = 1,\ldots,M, \quad (5.16)$$

which can also be written

$$p_{km} = \frac{\exp(\alpha_m - \beta c_{km} - \gamma \tilde{c}_m(\beta))}{\sum_{k=1}^{K}\sum_{m=1}^{M} \exp(\alpha_m - \beta c_{km} - \gamma \tilde{c}_m(\beta))},\ \ k = 1,\ldots,K,\ m = 1,\ldots,M. \quad (5.17)$$

Proof. The proposition follows from our general results (Proposition 3). □

This is the standard nested logit model (NL). The parameters α_m, β and γ have a direct coupling to the three constraints in the definition of cost-minimizing behavior. This gives a different explanation of what is at work in the model than the standard ARUM derivation that we will consider in Sect. 5.7.3.

Formula (5.17) can be written as a nested logit model,

$$p_{km} = \frac{\exp(\alpha_m - (\beta + \gamma)\tilde{c}_m(\beta))}{\sum_{m=1}^{M} \exp(\alpha_m - (\beta + \gamma)\tilde{c}_m(\beta))} \times \frac{\exp(-\beta c_{km})}{\sum_{k=1}^{K} \exp(-\beta c_{km})}. \quad (5.18)$$

Note that if the third inequality is removed from the definition we are back into the structured logit model with one cost constraint in Definition 12 and with the probability function

$$p_{km} = \exp(\mu + \alpha_m - \beta c_{km}),$$

which is (5.10), where β replaces γ.

We shall now consider the case with several cost constraints and a constraint on composite cost. If we in the definition have several cost measures (or attributes), one for each m, then this produces M parameters β_m, $m = 1, \ldots, M$, and the composite costs become $\tilde{c}_m(\beta_m) = -\frac{1}{\beta_m} \log \sum_{k=1}^{K} \exp(-\beta_m c_{km})$.

Definition 15 (Cost-Minimizing Behavior in the Standard Nested Logit Model with Several Cost Measures and a Constraint on Composite Cost). A probability distribution $p = (p_{km}, k = 1, \ldots, K, m = 1, \ldots, M)$ represents cost-minimizing behavior if and only if, for any sample size N,

$$\left.\begin{array}{l} \sum_{k=1}^{K} z_{km}^1 = \sum_{k=1}^{K} z_{km}^2, \; m = 1, \ldots, M, \\ \sum_{k=1}^{K} c_{km} z_{km}^1 \leq \sum_{k=1}^{K} c_{km} z_{km}^2, \; m = 1, \ldots, M, \\ \sum_{m=1}^{M} \tilde{c}_m(\beta_m) z_m^1 \leq \sum_{m=1}^{M} \tilde{c}_m(\beta_m) z_m^2, \end{array}\right\}$$
$$\Longrightarrow \prod_{k=1}^{K} \prod_{m=1}^{M} p_{km}^{z_{km}^1} \geq \prod_{k=1}^{K} \prod_{m=1}^{M} p_{km}^{z_{km}^2}.$$

The addition of the third inequality means that the two levels are not treated in a symmetric way. The choice of an alternative m in the upper level is governed not only by the costs c_{km} but by the composite cost $\tilde{c}_m(\beta_m)$ as well. In addition to the parameters α_m, $m = 1, \ldots, M$, and β_m, $m = 1, \ldots, M$ the third inequality generates an additional parameter γ.

Proposition 13 (Cost-Minimizing Behavior in the Standard Nested Logit Model with Several Cost Measures and a Constraint on Composite Cost). *The probability distribution $p = (p_{km}, k = 1, \ldots, K, m = 1, \ldots, M)$ represents cost-minimizing behavior with respect to the cost measures (c_{km}, $k = 1, \ldots, K$ and $m = 1, \ldots, M$) and the composite costs $\tilde{c}_m(\beta_m)$, $m = 1, \ldots, M$, if and only if there exist μ, α_m, $m = 1, \ldots, M$ and $\beta_m \geq 0$ $m = 1, \ldots, M$, and $\gamma \geq 0$ such that*

$$p_{km} = \exp(\mu + \alpha_m - \beta_m c_{km} - \gamma \tilde{c}_m(\beta_m)), \; k = 1, \ldots, K, \; m = 1, \ldots, M,$$

which can also be written

$$p_{km} = \frac{\exp(\alpha_m - \beta_m c_{km} - \gamma \tilde{c}_m(\beta_m))}{\sum_{k,m} \exp(\alpha_m - \beta_m c_{km} - \gamma \tilde{c}_m(\beta_m))}, \; k = 1, \ldots, K, \; m = 1, \ldots, M. \tag{5.19}$$

Proof. The proposition follows from our general results (Proposition 3). □

This is one version of the standard nested logit model (NL). The parameters α_m, β_m and γ have a direct coupling to the three constraints in the definition of cost-minimizing behavior. This gives a different explanation of what is at work in the model than the standard ARUM derivation that we will consider in Sect. 5.7.3.

The choice probabilities (5.19) can be written in the following equivalent form:

$$p_{km} = \frac{\exp(\alpha_m - \omega_m \tilde{c}_m(\beta_m))}{\sum_m \exp(\alpha_m - \omega_m \tilde{c}_m(\beta_m))} \times \frac{\exp(-\beta_m c_{km})}{\sum_k \exp(-\beta_m c_{km})}, \tag{5.20}$$

where $\omega_m = \beta_m + \gamma$.

This way of writing the choice probabilities indicates why it is called a "nested logit model". The choice probabilities p_{km} can be written as

$$p_{km} = p_m \times p_{k|m},$$

where p_m is the marginal probability of choosing alternative m in the upper level, and $p_{k|m}$ is the conditional probability of choosing alternative k in the lower level provided that alternative m is chosen for the upper level. In the standard nested model we can identify the factors in (5.20) as

$$p_m = \frac{\exp(\alpha_m - \omega_m \tilde{c}\, j_m(\beta_m))}{\sum_m \exp(\alpha_m - \omega_m \tilde{c}_m(\beta_m))},$$

and

$$p_{k|m} = \frac{\exp(-\beta_m c_{km})}{\sum_k \exp(-\beta_m c_{km})}.$$

The exponential, or in other words log linear, structure of logit models makes it easy to calculate marginal and conditional choice probabilities in this way. This can be repeated in a hierarchy of conditional probabilities. This mathematical property invites interpretations of the decision process as a stepwise procedure where decisions in the upper level are followed by decisions in the lower level. The behavioral motivation for this way of interpreting the decision process is unclear. In our formulation in Sect. 5.7.1 there is no immediate interpretation of such stepwise formulas.

Also, it is difficult to see the behavioral content of the introduction of cost measures expressed in terms of composite costs as in the standard nested logit formulation, Definitions 14 and 15. The structured logit models defined by Definitions 12 and 13 have a clear behavioral interpretation and seem to be preferable.

However, in the standard ARUM approach (see Sect. 5.7.3) in which the composite cost $\tilde{c}_m(\beta_m)$ represents the expected achieved perceived cost in the lower level the corresponding stepwise interpretation is often made. However, the behavioral content of this approach is dubious, since it is hard to see how the decision maker would have the capacity to calculate the composite cost before making the decision.

The composite cost has another interpretation as we shall see in Chap. 6: the advantage/welfare of the choices on the lower level, which includes costs and freedom of choice.

5.7.3 The Standard ARUM Nested Approach

In the ARUM approach the assumption of independent extreme value distributed random variables can be relaxed in order to permit dependent random variables by using Generalized Extreme Value distributed random components (Sect. 4.5.3). This offers one way of deriving the standard nested logit model.

The choice probabilities can be obtained by specifying the generating function G (4.13) in the following way:

$$G(X_{11}, \ldots, X_{km}, \ldots, X_{KM}) = \sum_{m=1}^{M} \exp(\alpha_m)(\sum_{k=1}^{K} X_{km}^{\beta_m})^{\frac{1}{\beta_m}}.$$

According to Proposition 4 the choice probabilities are given by

$$p_{km} = \frac{\exp(-\theta c_{km}) G^{km}(\exp(-\theta c_{11}), \ldots, \exp(-\theta c_{KM}))}{G(\exp(-\theta c_{11}), \ldots, \exp(-\theta c_{KM}))},$$

where G^{km} is the (km)-th partial derivative of G with respect to X_{km} and θ is the parameter of the generalized extreme value distribution and the utilities v_k have been replaced with negative costs $-c_{km}$. Hence,

$$p_{km} = \frac{\exp(-\theta c_{km}) \exp(\alpha_m)(\sum_{k=1}^{K} \exp(-\beta_m \theta c_{km}))^{\frac{1}{\beta_m}-1} \exp(-\theta(\beta_m - 1)c_{km})}{\sum_{m=1}^{M} \exp(\alpha_m)(\sum_{k=1}^{K} \exp(-\beta_m \theta c_{km}))^{\frac{1}{\beta_m}}}.$$

By writing $\gamma_m = \theta\beta_m$, and using the composite cost we obtain

$$p_{km} = \frac{\exp(\alpha_m - (\gamma_m/\theta)\tilde{c}_m(\gamma_m))}{\sum_m \exp(\alpha_m - (\gamma_m/\theta)\tilde{c}_m(\gamma_m))} \times \frac{\exp(-\gamma_m c_{km})}{\sum_k \exp(-\gamma_m c_{km})}.$$

This is (5.20) if we identify $\beta_m = \gamma_m$ and $\omega_m = \gamma_m/\theta$. Hence, the standard nested ARUM model is identical with the cost-minimizing model derived from Definition 15.

5.8 The Logit Model with Individual Cost Values

So far we have implicitly assumed that the cost of each alternative is the same for each decision maker. However, our derivation can easily be extended to the case where there is individual variation in the cost values. In Sect. 4.6.3 we derived the choice probability distribution for this case.

We shall now discuss the situation where each individual decision maker is characterized by a (possibly) different cost of the alternatives. We shall assume that each decision maker has cost-minimizing behavior according to Definition 7 so that the corresponding probability distribution is given by Proposition 6.

Consider a population/sample of M decision makers. Let the cost of alternative k for decision maker number m be c_k^m, $k = 1, \ldots, K$, and $m = 1, \ldots, M$. Then, if the decision makers are cost minimizing, the choice probabilities are given by

$$p_k^m = \frac{\exp(-\gamma c_k^m)}{\sum_{k=1}^K \exp(-\gamma c_k^m)},$$

where p_k^m denotes the probability distribution for decision maker number m. Here we have assumed that the parameter γ is the same for all decision makers. Introduce the variables

$$y_k^m = \begin{cases} 1 \text{ if the decision maker } m \text{ chooses alternative } k, \\ 0 \text{ otherwise.} \end{cases}$$

Let d denote the decisions by the M decision makers. Then the probability of the decisions d can be written

$$\begin{aligned} \Pr(d) &= \prod_{m=1}^M \prod_{k=1}^K (p_k^m)^{y_k^m} = \prod_{m=1}^M \prod_{k=1}^K \left(\frac{\exp(-\gamma c_k^m)}{\sum_{k=1}^K \exp(-\gamma c_k^m)}\right)^{y_k^m} \\ &= \frac{\exp(-\gamma \sum_{m=1}^M \sum_{k=1}^K c_k^m y_k^m)}{\prod_{m=1}^M \prod_{k=1}^K (\sum_{k=1}^K \exp(-\gamma c_k^m))^{y_k^m}}. \end{aligned}$$

5.9 Socioeconomic Factors

So far we have shown how to derive models intended to take care of influences by costs and other quantitative attributes on the decision making. We shall now see how qualitative individual attributes such as sex, age groups and income classes can be handled. Let the decision makers be classified into S income classes, $s = 1, \ldots, S$. We wish to take care of the fact that cost-minimizing behavior may differ between income classes. Hence, in defining cost-minimizing behavior we wish to compare samples with the same number of decision makers in the classes. We formulate this by using activity equivalence. Let the number of decision makers belonging to class s choosing alternative k with cost c_{sk} be z_{sk}. Note that the cost c_{sk} may be a generalized cost including time costs, in which case decision makers may value time differently depending on which income class they belong to. As before, let there be two samples/samples denoted z_{sk}^1 and z_{sk}^2. Activity equivalence will be defined by the second condition in the definition below.

Definition 16 (Cost-Minimizing Behavior and Socioeconomic Factors, Case I). A probability distribution $p = (p_{sk}, s = 1, \ldots, S, k = 1, \ldots, K,)$ represents cost-minimizing behavior if and only if, for any sample size N,

$$[\sum_{k=1}^{K} c_{sk} z_{sk}^1 \leq \sum_{k=1}^{K} c_{sk} z_{sk}^2 \text{ and } \sum_{k=1}^{K} z_{sk}^1 = \sum_{k=1}^{K} z_{sk}^2, \ s = 1, \dots, S]$$
$$\Longrightarrow \prod_{s=1}^{S} \prod_{k=1}^{K} p_{sk}^{z_{sk}^1} \geq \prod_{s=1}^{S} \prod_{k=1}^{K} p_{sk}^{z_{sk}^2}.$$

Here activity equivalence is defined to hold if the number of decision makers in each income group s is the same in both samples. Cost-minimizing behavior obtains if the total income for each income group is not larger in the first sample than in the second sample, provided that the samples are activty equivalent.

Proposition 14 (Cost-Minimizing Behavior and Socioeconomic Factors, Case I). *The probability distribution $p = (p_{sk}, \ s = 1, \dots, S, \ k = 1, \dots, K,)$ represents cost-minimizing behavior if and only if there exists a nonnegative vector $\gamma \in \mathrm{R}^S$, such that for some vector $\alpha \in \mathrm{R}^M$, and some $\mu \in \mathrm{R}$,*

$$p_{sk} = \exp(\mu + \alpha_s - \sum_{s=1}^{S} \gamma_s c_{sk}),$$

or equivalently

$$p_{sk} = \frac{\exp(\alpha_s - \sum_{s=1}^{S} \gamma_s c_{sk})}{\sum_{k=1}^{K} \sum_{s=1}^{S} \exp(\alpha_s - \sum_{s=1}^{S} \gamma_s c_{sk})}. \tag{5.21}$$

Proof. The proposition follows from our general results, Proposition 3.
To see this take A equal to the $S \times K$ matrix

$$A = \begin{bmatrix} 11\cdots1 & 0 & 0 & \cdots & 0 \\ 0 & 11\dots1 & 0 & \cdots & 0 \\ 0 & 0 & 11\cdots1 & & 0 \\ \vdots & & & \ddots & \vdots \\ 0 & 0 & & \cdots & 11\cdots1 \end{bmatrix},$$

and $z = (z_{11}, z_{12}, \dots, z_{1K}, z_{2,1}, \dots, z_{2K}, \dots, z_{S1}, \dots, z_{SK})$ in Proposition 3. Note that s corresponds to m and z_{sk} corresponds to z_k in Proposition 3. Note furthermore that c_{sk} are the elements of the matrix C. □

A special case of the previous situation obtains if costs do not depend on income class, $c_{sk} = c_k, \ s = 1, \dots, S$. The S cost conditions in the definition reduces to one condition.

Definition 17 (Cost-Minimizing Behavior and Socioeconomic Factors, Case II). A probability distribution $p = (p_{sk}, \ s = 1, \dots, S, \ k = 1, \dots, K,)$ represents cost-minimizing behavior if and only if, for any sample size N,

$$[\sum_{k=1}^{K}\sum_{s=1}^{S} c_k z_{sk}^1 \leq \sum_{k=1}^{K}\sum_{s=1}^{S} c_k z_{sk}^2 \text{ and } \sum_{k=1}^{K} z_{sk}^1 = \sum_{k=1}^{K} z_{sk}^2,\ s = 1, \dots, S]$$
$$\Longrightarrow \prod_{s=1}^{S}\prod_{k=1}^{K} p_{sk}^{z_{sk}^1} \geq \prod_{s=1}^{S}\prod_{k=1}^{K} p_{sk}^{z_{sk}^2}.$$

We obtain:

Proposition 15 (Cost-Minimizing Behavior and Socioeconomic Factors, Case II). *The probability distribution* $p = (p_{sk},\ s = 1, \dots, S,\ k = 1, \dots, K,)$ *represents cost-minimizing behavior if and only if there exists a nonnegative* $\gamma \in \mathrm{R}$, *such that for some vector* $\alpha \in \mathrm{R}^M$, *and some* $\mu \in \mathrm{R}$,

$$p_{sk} = \exp(\mu + \alpha_s - \gamma c_k),$$

or equivalently

$$p_{sk} = \frac{\exp(\alpha_s - \gamma c_k)}{\sum_{k=1}^{K}\sum_{s=1}^{S} \exp(\alpha_s - \gamma c_k)}.$$

By introducing the zero/one indicator variables

$$\delta_{sk} = \begin{cases} 1 & \text{if the decision maker belongs to income group } s, \\ 0 & \text{otherwise,} \end{cases}$$

formula (5.21) can be written

$$p_{sk} = \frac{\exp(\sum_{s=1}^{S} \delta_{sk}(\alpha_s - \gamma_s c_{sk}))}{\sum_{k=1}^{K}\sum_{s=1}^{S} \exp(\sum_{s=1}^{S} \delta_{sk}(\alpha_s - \gamma_s c_{sk}))}.$$

5.10 Comments and Extensions

In deriving the multi-attribute/linear-in-parameters multinomial logit model (5.2) by the classic random utility approach the Gumbel parameter, here denoted by μ, appears as a factor in the exponential:

$$p_k = \frac{\exp(-\mu \sum_{s=1}^{S} \beta_s c_{sk})}{\sum_{k=1}^{K} \exp(-\mu \sum_{s=1}^{S} \beta_s c_{sk})},\quad k = 1, \dots, K.$$

The value of the parameter μ is often considered to be arbitrary and therefore often set to 1 (See e.g. Ben-Akiva and Lerman 1985). However, this may be misleading if the standard interpretation of the composite utility as the expected achieved utility is being used. A thorough discussion is given by Brundell-Freij (1995).

Nested (structured) logit models have been the common approach in dealing with different decision situations in transportation, such as choice of mode, choice of trip origin, trip destination and choice of route. The standard nested logit model should not be looked upon as a model of behavior. The hierarchical structure of the mathematical formulas invites a stepwise interpretation of the decision process. However, the hierarchical structure should be looked upon as a conceptual structure only in a situation "where we might wish to regard the choices contained in a particular subset of the total choice set as having more in common with each other than with the remaining choices" (Daly and Zachary 1978). If a stepwise interpretation is made using the composite cost/log sum with the interpretation of expected achieved utility in the lower level it is difficult to see how the trip makers would be able to estimate this complex function. It may be difficult enough for the trip makers to estimate the values of the cost attributes. A more reasonable approach seems to be to use the structured logit model given by (5.11) and (5.13). In this formulation Definitions 12 and 13 brings out exactly what is assumed and which alternatives are grouped together. Furthermore it gives an explanation of the function of the particular upper level parameters (α_m) which otherwise often are introduced in an ad hoc manner to take care of hidden differences between the alternatives. Ben-Akiva introduces alternative-specific constants and alternative-specific socioeconomic variables. These are also called dummy variables, they take values of 0 or 1 only (Ben-Akiva and Lerman 1985, p. 75–77).

The standard nested logit model (5.17) and (5.18) has the additional parameter γ. The addition of parameters may accidentally give a better fit to data when estimating the parameters even if the behavioral interpretation is unclear. However, the prediction power may be poor if the components of the mathematical model does not correspond to behavior. Hence the prediction power of the standard nested logit model may still be poor if the model does not correspond to behavior even if it contains one additional parameter. The underlying behavioral assumptions of the standard nested logit model (5.18) derived within the ARUM approach cannot be tested since the underlying random variables (X_k of Sect. 4.5.3) cannot be observed, and thus the agreement with observations cannot be tested. The derived choice probabilities can be tested, but not the underlying assumptions. However, the assumptions made in our cost-minimizing behavior approach in Definitions 14 and 15 in Sect. 5.7.2 can be tested. Hence our approach, leading to the standard nested logit model by a different formulation of the assumptions, offers a way to investigate the standard nested logit model.

The derivation of nested logit models from Generalized Extreme Value distributed random components (Sect. 5.7.3) is a mathematical exercise that does not give any insight into what is at work from a behavioral perspective. Also it has the same difficulties as the derivation by using extreme value distributed components, since the underlying random variables cannot be observed. Hence the agreement between the assumptions of the model and data cannot be investigated. Only derived consequences of the model such as the choice probability function can be tested.

A mixed logit model is a logit model with random coefficients. When the cumulative distribution function describing the random variation has finite support we talk about latent class models. We do not proceed into this direction except for noting that any parameter/coefficient in the logit models treated in this book can be looked upon as a random variable. Any logit model can be extended into a mixed logit model or a latent class model, provided we are willing to make the necessary assumptions about the unknown mixing probability distribution.

5.11 Notes

For the stochastic route choice model, Sect. 5.2, see e.g., Daganzo and Sheffi (1977), Sheffi (1985), Bell and Iida (1997).

The multi-attribute/linear-in-parameters multinomial logit model, Sect. 5.3, is one of the standard models (see e.g. McFadden 1974). Ben-Akiva and Lerman (1985), p. 108, calls this Linear-in-Parameters Logit Model. The section is based on Erlander (2002). The gravity model, Sect. 5.5, has a long history, see e.g. Erlander and Stewart (1990). It was derived from efficiency by Smith (1978a). Smith gives a comprehensive detailed overview over axiomatic derivations of gravity models in his investigation of these models (Sen and Smith 1995). The structured logit model, Sect. 5.7.1, is the Joint Logit model (Ben-Akiva and Lerman 1985), whereas the standard ARUM nested logit model, Sect. 5.7.3, is the Nested Logit Model (Ben-Akiva and Lerman 1985). The standard nested logit model, Sect. 5.7.3, was developed within the ARUM approach by Ben-Akiva (1974), Williams (1977) ((5.18) is essentially formula (158) in Williams 1977), Williams and Senior (1977), Daly and Zachary (1978) and McFadden (1978a). For an overview see Ortuzar (2001). For a discussion of socio-economic factors see e.g. Erlander and Stewart (1990). For an overview of mixed logit models see McFadden and Train (2000). See also Greene and Hensher (2003).

Chapter 6
Welfare, Benefit and Freedom of Choice

Abstract A distinction is made between advantage and achievement. The observed entropy of a sample is used as a measure of *revealed freedom of choice*. A *measure of welfare* (advantage) is obtained by combining negative expected cost (achievement) and entropy (freedom of choice). The *measure of welfare* (advantage) turns out to be the so called *composite cost*.

6.1 Introduction

Demand for travel depends on the localization of housing and work places and other centers of activities. The network users make choices with regard to route, travel mode, origin and destination and the traffic conditions. Factors influencing the amount of travel and the way the traffic system is being used are money costs, time costs, parking costs etc. Factors such as fare policies, traffic control policies and the construction of new roads also influence the choices.

The minimum cost would be obtained if every trip maker chooses the cheapest alternative. But this is not the interesting case. Rather, the case of interest is when there is some variation in the choices. This variation can be given the interpretation of freedom of choice. Large variation means more freedom of choice. We shall show how a measure of freedom of choice can be constructed and how this freedom of choice measure can be combined with average cost to give a welfare measure or as we will call it a measure of advantage. Freedom of choice can also be called degree of choice.

We shall distinguish between two things: the *advantage* to the network users and the *achievements* of the network users. The distinction between advantage and achievement was introduced by Amartya Sen in his discussion of freedom of choice (Sen 1988).

The achievement of the network users will be measured by the negative of average cost. By using the negative of average cost as a measure of achievement of the network users the traffic planner can analyze and evaluate the effects of changes in management, control, design and the effects of investments aiming at the improvement of the performance of the system. However this is not enough, changes in

S. Erlander, *Cost-Minimizing Choice Behavior in Transportation Planning*, Advances in Spatial Science, DOI 10.1007/978-3-642-11911-8_6,

the freedom of choice has also to be taken account of. Consider, for example, the trip distribution problem treated in Sect. 5.5. The minimum average cost would be obtained if the network users (were forced to) follow a linear programming optimal solution, resulting in less freedom of choice. But this is usually far from the behavior that can be observed in practice. Hence, we need a measure of freedom of choice in order to evaluate and compare different plans for the development of the traffic system.

Freedom of choice, as we use the notion, is a property of the trip pattern obtained when the trip makers have made their choices. Freedom of choice is related to the quality of the traffic system and as such it can be used by the traffic planner to evaluate the traffic system and compare different plans for the development of the traffic system. The presumption is that the choices made by network users relate to factors such as money costs, time costs, parking costs, fare policies, traffic control policies etc. Freedom of choice is not a factor considered by the trip maker in the decision process.

We shall introduce a measure of freedom of choice in Sect. 6.4 and then in Sect. 6.5 combine this freedom of choice measure with the measure of achievement discussed above to obtain a measure of welfare or advantage, which includes the achievements. However, we shall begin by discussing freedom of choice detached from the choice mechanism; aiming at a freedom of choice measure that can be used to compare different realized trip patterns irrespective of the mechanism behind the occurrence of specific trip patterns.

In classic discrete choice theory, ARUM theory, composite utility – the so called log sum – has the interpretation of expected achieved perceived utility. In microeconomics it therefore has the interpretation of indirect utility. It is also the consumer surplus function. We shall see that our measure of advantage is equal to the composite utility, thus giving a new interpretation of this quantity.

6.2 Achievement Measure

We shall begin by repeating some notation. Let Ω be the set of all decision makers. The cost of the alternatives are given by c_k, $k = 1, \ldots, K$. Let $p_k = \Pr(d_n = k)$ denote the probability that a decision maker drawn at random from the population Ω chooses alternative k, $k = 1, \ldots, K$. Let $d = (d_1, \ldots, d_N)$ and $z(d) = (z_1(d), \ldots, z_K(d))$ denote the decision patterns and the frequencies respectively in a finite sample of decision makers – trip makers – of size N.

The benefit - or achievement as we shall call it - of the decision makers in a finite sample of size N is measured by the negative average cost $-\bar{c}(d) = -\frac{1}{N}\sum_{k=1}^{K} c_k z_k(d)$. Taking expected values we obtain $E[-\bar{c}(d)] = -\sum_{k=1}^{K} c_k p_k = -c$.

In the multi attribute logit model, (5.2), we have to translate the different costs or attributes into say, units of the first cost. This is done by using the rates of substitution β_s/β_1 to obtain the generalized cost $c_k = \sum_{s=1}^{S}(\beta_s/\beta_1)c_{sk}$ (see Sect. 5.4).

Then the average negative generalized cost becomes

$$-\bar{c}(d) = -\frac{1}{N}\sum_{k=1}^{K} c_k z_k(d) = -\sum_{k=1}^{K}\sum_{s=1}^{S}(\beta_s/\beta_1)c_{sk}\frac{z_k(d)}{N}$$

with expected value equal to

$$E[-\bar{c}(d)] = -\sum_{k=1}^{K} c_k p_k = -\sum_{k=1}^{K}\sum_{s=1}^{S}(\beta_s/\beta_1)c_{sk}p_k.$$

6.3 Freedom of Choice: Preliminaries

What can be concluded from the choices made? Can the choices made by the decision makers in a decision pattern d tell us something about the freedom of choice? What does the observed decisions reveal about the freedom of choice? Does it make sense to construct an index of freedom of choice comparing freedom of choice in one decision pattern d^1 with freedom of choice in another decision pattern d^2?

Freedom of choice could be related to the choice set. The simplest measure of freedom of choice would be the cardinality K of the choice set. However, this does not give much information when the planner is trying to compare different plans for the development of the road system.

Giving a decision maker more choice by adding alternatives might be valuable. Surely, adding alternatives would not decrease freedom of choice. Will adding of alternatives always increase freedom of choice? Not necessarily. Adding an alternative with very high cost will not increase freedom of choice if the alternative is never chosen.

So this leads up to the conclusion that we want a measure of freedom of choice which is based on the choice set but which also somehow takes into account the costs of the alternatives and other constraints acting on the choices. One way of doing this is to make use of the information contained in the actual choices made by the decision makers. The decisions taken by the decision makers reflect available alternatives, costs of the alternatives and other constraints. The idea that we shall use is simply to observe the decisions $d = (d_1, \dots, d_N)$ made by a random sample of N decision makers resulting in the frequencies $z(d) = (z_1(d), \dots, z_K(d))$. Then trough a combinatorial argument we construct a measure of freedom of choice based on the number of ways of obtaining the particular frequency vector $z(d)$. The freedom of choice is high if the observed frequency vector can be obtained in many ways. In this way the facts governing the decision making is indirectly taken care of.

We shall begin with a less formal discussion of a route choice problem. Each one of N trip makers makes a decision about which route to take. The resulting decision pattern d reflects many factors. An important factor considered by the trip makers is the (generalized) cost of choosing a specific route. Trip makers with other origins and destinations compete about the existing road network. In a broader context, the number of trip makers going from an origin to a destination may be constrained by the localization of living sites and work places. Links in the road network may operate close to capacity limits. Hence, decisions by the trip makers may be constrained by many factors, explicit and implicit. Some factors may be included in the (generalized) cost, others not.

Consider the following small example. Let there be $N = 10$ trip makers choosing between two routes r_1 and r_2. The total number of feasible trip patterns is $2^{10} = 1024$. Assume that there are capacity constraints, such that not more than 2 trip makers can take route r_1, at most 8 trip makers can take route r_2. If $z_k(d)$ is the number of trip makers choosing route k, then the frequencies have to be $z_1(d) = 2$ and $z_2(d) = 8$. The first trip maker can choose between r_1 and r_2. The same is true for the second trip maker. The following trip makers have the same choice situation as long as the capacity of r_1 is not reached. As soon as two trip makers have chosen route number r_1, the other remaining eight trip makers have to choose route number r_2 for the frequencies $(2, 8)$ to obtain. So even if there are strong capacity constraints there remains some freedom of choice for the individual trip makers. The frequencies $(2, 8)$ can be obtained in $(10!/2!8!) = 45$ ways. This number can be taken as a freedom of choice measure; it tells the number of trip patterns d that gives the frequencies (2,8). This freedom of choice measure takes the value 45 for every trip pattern d giving the frequencies (2,8). Let now the capacity constraints be removed. Assume that for some reason a trip pattern d giving the previous frequencies $(2, 8)$ is obtained. Then the same value of the freedom of choice measure is obtained for this trip pattern. Let now the frequencies (3,7) be obtained for another trip pattern, giving the freedom of choice value $(10!/3!7!)= 120$. The latter trip pattern can be obtained in 120 ways and there is more freedom of choice. The highest value of the freedom of choice measure is obtained for the frequencies (5,5), which gives $(10!/5!5!) = 252$. Let now a third route r_3 be added. The total number of feasible trip patterns is now $3^{10} = 59049$. Again, the frequencies $(2, 8, 0)$ give the freedom of choice measure $(10!/2!8!0!) = 45$, whereas e.g. $(2, 7, 1)$ gives $(10!/2!7!1!) = 360$. The greatest freedom of choice is obtained for $(3, 3, 4)$ with value $(10!/3!3!4!) = 4200$.

Assume now that three more routes are added. We might think that freedom of choice is then increased. However, if the three additional routes are never chosen, the situation has not changed, and the same freedom of choice will attain. The greatest freedom of choice will be the same as before for the frequencies $(3, 3, 4, 0, 0, 0)$ with value $(10!/3!3!4!0!0!0!) = (10!/3!3!4!) = 4200$. This situation will often occur in practical applications, since in a large network there will be many routes that are never chosen because of their length or travel time.

Note that the (generalized) cost is determined by the frequencies $z_k(d)$, only: $c(d) = \sum_{k=1}^{K} c_k z_k(d)$. Hence the frequencies $z_k(d)$ determine the (generalized) cost as well as the value of the freedom of choice measure above.

The small example above clearly demonstrates that the combinatorial formula used is a good measure of the degree of freedom related to different trip patterns. This will be the starting point for the freedom of choice measure that we are now ready to introduce in the next section. It has been suggested that the number of available routes could be used as a freedom of choice measure. However, such a freedom of choice measure is not related to the utilization of the traffic system and is therefore of limited value for our purpose.

6.4 Freedom of Choice Measure

Let there be N decision makers, each making one decision, i.e. choosing one alternative from the choice set containing K alternatives. The decision pattern $d = (d_1, \ldots, d_n, \ldots, d_N)$ denotes the simultaneous decisions by all N decision makers. There is a finite number of decision patterns. Hence, the decision patterns can be enumerated, $d^1, \ldots, d^{K_N}$, where K_N is a usually very large number. The simultaneous choices by the N trip makers give rise to a specific trip pattern d. Hence, we can talk about the *simultaneous choice set* $\mathcal{D} = [d^1, \ldots, d^{K_N}]$. The simultaneous decisions by the N decision makers can be described as the choice of one particular element $d \in \mathcal{D}$. Many elements of $\mathcal{D}$ are similar with respect to one or more properties. E.g., all permutations of the order of the decisions in a decision pattern d give decision patterns containing the same subset of decisions. It would make sense to say that the *revealed freedom of choice* is the same for all elements of such a subset. In particular, we shall use the combinatorial (6.1) below to obtain a partition of the set $\mathcal{D}$ into equivalence classes having the same freedom of choice.

We shall formulate a freedom of choice measure based on the number of possible decision patterns d resulting in a particular frequency vector $z = z(d) = (z_1(d), \ldots, z_K(d)), z : Z^N_{++} \to Z^K_+$. The number of ways to obtain the frequency vector z is a freedom of choice measure. We shall see that by approximating by Stirling's formula the combinatorial formula given below, neglecting constant terms, taking logarithms and dividing by the sample size we arrive at the observed entropy as a freedom of choice measure.

There will usually be many decision patterns d that result in a particular vector $z = z(d)$. Define the inverse image $z^{-1}(z) : Z^K_+ \to Z^N_{++}$,

$$z^{-1}(z) = D_z = [d \mid z(d) = z, z = (z_1, \ldots, z_K)].$$

The inverse image $z^{-1}(z) = D_z$ partitions the simultaneous choice set $\mathcal{D}$ into equivalence classes giving rise to the same frequency vector z. The number of elements in the set D_z, the cardinality of D_z, $K(D_z)$, is given by

$$K(D_z) = \frac{N!}{\prod_{k=1}^K z_k!}. \tag{6.1}$$

The decision makers can choose any decision pattern belonging to the set D_z for $z = z(d)$ to obtain, $d \in D_{z(d)}$. The cardinality of $D_{z(d)}$ is a measure of the freedom of choice in choosing d. If the cardinality is one, i.e. there is just one element in the set $D_{z(d)}$, say $d = (1, 1, \ldots, 1)$, then all the decision makers have to choose alternative 1; their freedom of choice is small. On the other hand, if the cardinality is large, then the decision makers have a large freedom of choice, since there are many decision patterns in the set $D_{z(d)}$ which are compatible with the value of $z(d)$; the decision makers has to choose alternative k $z_k(d)$ times, $k = 1, \ldots, K$, but is otherwise not restricted.

Note that our purpose is to define a measure of freedom of choice. This measure will be the same irrespective of the process used by the decision makers in making the decision.

We could take the cardinality of $D_{z(d)}$ as our measure of freedom of choice. However, we can equally well define our measure of freedom of choice by taking a continuous monotone function of the cardinality. Define the freedom of choice measure Φ by

$$\Phi(z) = \frac{1}{N} \log K(D_z) = \frac{1}{N} \log K(z^{-1}(z)) = \frac{1}{N} \log \frac{N!}{\prod_{k=1}^{K} z_k!}.$$

We have here defined our measure of freedom of choice in terms of z alone, since the value will be the same for all $d \in D_z$. By using Stirling's formula, neglecting constant terms, $\Phi(z)$ can be written approximately,

$$\begin{aligned} \Phi(z) &\approx \frac{1}{N}(-N + N \log N - (-N + N \sum_{k=1}^{K} (z_k/N) \log((z_k/N)N))) \\ &= - \sum_{k=1}^{K} (z_k/N) \log(z_k/N) = H([z_k/N]), \end{aligned} \tag{6.2}$$

where $H()$ is the entropy function. Hence, a natural measure of the freedom of choice is the entropy $H([z_k/N])$; this is the observed entropy of the sample, if we think of d as a sample. The observed entropy (6.2) will be called *the revealed freedom of choice.* It is (approximately) the logarithm of the number of ways to obtain the sample divided by the sample size N. A high value corresponds to many ways of obtaining the sample, a low value the opposite.

- We shall take the *observed entropy*, $H([z_k/N])$, as our *measure of revealed freedom of choice.*

So far in this section we have not specified the decision process. The decision process could be based on preferences or utilities or be specified in some other way. It could be deterministic or stochastic. In many situations it is natural to assume that the decision process is stochastic. By describing the decision process probabilistically many decision processes can be included in an approximate way, even though we do not specify the details.

6.4.1 Freedom of Choice in the Probabilistic Case

We shall now return to the case where we describe the decision making by a choice probability function. Recall that p_k is the probability of choosing alternative $k, k = 1, \ldots, K$, $p = (p_1, \ldots, p_K)$ and let $\bar{p}_k = z_k/N$, $\bar{p} = (\bar{p}_1, \ldots, \bar{p}_K)$. Here $\bar{p}$, the relative frequencies, is the maximum likelihood estimate of the probability distribution p, if we do not know the structure of p. We have

$$-H(\bar{p}) = \sum_{k=1}^{K} (z_k/N) \log(z_k/N) = \sum_{k=1}^{K} (z_k/N) \log p_k (z_k/Np_k)$$
$$= \sum_{k=1}^{K} (z_k/N) \log p_k + \frac{1}{N} \sum_{k=1}^{K} z_k \log(z_k/Np_k). \tag{6.3}$$

The last term can be approximated by

$$\sum_{k=1}^{K} z_k \log(z_k/Np_k) \approx \frac{1}{2} \sum_{k=1}^{K} \frac{(z_k - Np_k)^2}{Np_k},$$

and we obtain,

$$-H(\bar{p}) \approx \sum_{k=1}^{K} (z_k/N) \log p_k + \frac{1}{2N} \sum_{k=1}^{K} \frac{(z_k - Np_k)^2}{Np_k}.$$

The observed frequencies $[z_k]$ have a multinomial probability distribution, and hence, for large N, z_k is approximately normal with expected value $= Np_k$ and variance $= Np_k(1 - p_k) \approx Np_k$. Hence, for large N,

$$\sum_{k=1}^{K} \frac{(z_k - Np_k)^2}{Np_k}$$

has a χ^2-distribution with $(K - 1)$ degrees of freedom since there is the linear relation $\sum_{k=1}^{K} z_k = N$. Then the expected freedom of choice is given by

$$E[\Phi(z)] \approx E[H([z_k/N])] = E[H(\bar{p})] \approx H(p) - \frac{K-1}{2N}.$$

For large N the expected freedom of choice will be approximately $H(p)$. Hence, the entropy function, $H(p)$, is a natural measure of the expected freedom of choice in relation to the choice process described by the probability function p over the choice set of alternatives, $k = 1, \ldots, K$.

- We shall take the *entropy*, $H(p)$, as our *measure of expected freedom of choice*.

The notion of entropy is discussed in the Appendix, Sects. 9.2 and 9.3.

6.5 Welfare Measures

We shall now return to the discussion of welfare and benefit measures. In Sect. 6.1 we found that the welfare contribution from the transportation system had to include the advantage of freedom of choice as well as the achievement of low costs.

The achievement of the decision makers in a finite sample of size N can be measured by the negative average cost $-\bar{c} = -\sum_{k=1}^{K} c_k z_k / N$. Revealed freedom of choice can be measured by the observed entropy of the sample

$$H(\bar{p}) = -\sum_{k=1}^{K} (z_k(d)/N) \log(z_k(d)/N).$$

The pair $(-\bar{c}(d), H(\bar{p}))$ is a two-dimensional measure of advantage or welfare based on cost and freedom of choice. We shall now see how, due to the special form of the choice probability functions in the logit case, this two-dimensional advantage measure can be condensed into one one-dimensional measure.

6.5.1 *Welfare Measure for the Simple Logit Model*

The logit model has the choice probabilities

$$p_k = \frac{\exp(-\beta c_k)}{\sum_{k=1}^{K} \exp(-\beta c_k)}, k = 1, \ldots, K. \tag{6.4}$$

From this and (6.3) follows that minus the measure of revealed freedom of choice can be written

$$\begin{aligned} -H(\bar{p}) \approx &= \sum_{k=1}^{K} (z_k/N) \log p_k + \frac{1}{N} \sum_{k=1}^{K} z_k \log(z_k/N p_k) \\ &= -\beta \sum_{k=1}^{K} c_k (z_k/N) - \log \sum_{k=1}^{K} \exp(-\beta c_k) + \frac{1}{N} \sum_{k=1}^{K} z_k \log(z_k/N p_k) \\ &= -\beta \bar{c} - \log \sum_{k=1}^{K} \exp(-\beta c_k) + \frac{1}{N} \sum_{k=1}^{K} z_k \log(z_k/N p_k), \end{aligned}$$

where $\bar{c} = \sum_{k=1}^{K} c_k (z_k/N)$. From this expression it can be seen that $H(\bar{p})/\beta$ has the same dimensional property as average cost $\bar{c}$. Hence the observed entropy divided by the parameter β is measured in the same terms as average cost. This means that the components of the two-dimensional measure of advantage or welfare $(-\bar{c}, H(\bar{p}))$ can be combined into one one-dimensional measure by adding together

$-\bar{c}$ and $H(\bar{p})/\beta$,

$$W^{logit}(\bar{p}) = -\bar{c} + \frac{1}{\beta}H(\bar{p}),$$

where we have taken negative cost to change cost into advantage. We give this as a proposition.

Proposition 16. *For the logit model the observed combined advantage $W^{logit}(\bar{p})$ of negative average cost and revealed freedom of choice for the decision makers is equal to*

$$W^{logit}(\bar{p}) = -\bar{c} + \frac{1}{\beta}H(\bar{p}).$$

We have thus obtained a one-dimensional advantage measure for a sample from the logit model.

The corresponding result for the probability distribution itself will now be obtained. In the logit model the parameter β (and the given values $c_1, \ldots, c_K$, of course) completely determines the probabilities, and hence also the values of $c = E[\bar{c}] = \sum_{k=1}^{K} c_k p_k$ and $H(p) = -\sum_{k=1}^{K} p_k \log p_k$. For the logit model the entropy can be written

$$\begin{aligned} -H(p) &= \sum_{k=1}^{K} p_k \log p_k \\ &= -\beta \sum_{k=1}^{K} p_k c_k - \log \sum_{k=1}^{K} \exp(-\beta c_k) = -\beta(c - \tilde{c}(\beta)), \end{aligned} \tag{6.5}$$

where $\tilde{c}(\beta)$ is the composite cost defined earlier by (4.2),

$$\tilde{c}(\beta) = -\frac{1}{\beta} \log \sum_{k=1}^{K} \exp(-\beta c_k). \tag{6.6}$$

From (6.5) it can be seen that $H(p)/\beta$ has the same dimensional property as (generalized) cost c. Hence entropy divided by the parameter β is measured in the same terms as cost. But this means that we can once more combine cost and entropy into one one-dimensional measure,

$$W^{logit}(p) = -c + \frac{1}{\beta}H(p), \tag{6.7}$$

where we have taken negative cost to change cost into advantage. Combining (6.7) with (6.5) we obtain the proposition below:

Proposition 17. *For the logit model the combined advantage $W^{logit}(p)$ of expected negative cost and expected freedom of choice for the decision makers is equal to the* *negative of composite cost*

$$W^{logit}(p) = -c + \frac{1}{\beta} H(p) = -\tilde{c}(\beta). \tag{6.8}$$

Proposition 17 can also be obtained from the fact that a first degree approximation of changes in entropy as a function of changes in cost can be written

$$\Delta H \approx \frac{\delta H}{\delta c} \Delta c = \beta \Delta c,$$

which means that changes in entropy can be translated into changes in cost by dividing by β. Since the entropy $H(p) \geq 0$ it follows from (6.8) that

$$\tilde{c}(\beta) \leq c.$$

From (6.8) it can be seen that the composite cost $\tilde{c}(\beta)$ decreases without bounds if the parameter β becomes small. This reflects the fact that for small values of β the cost values c_k contribute little to the results. Instead, as the choice probabilities become more equal as $\beta \to 0$, the freedom of choice as measured by $(1/\beta)H(p)$ increases without bounds. Hence, if we interpret the negative of the composite cost, $-\tilde{c}(\beta)$, as a measure of advantage including freedom of choice, then the behavior for small values of the parameter β agrees with what our intuition tells. Note that $W^{logit}(p)$ grows without bounds as β becomes small, because we measure freedom of choice in terms of cost. The entropy itself attains the value $\max H(p) = \log K$, as $\beta \to 0$. On the other hand if the parameter β grows without bounds, $\beta \to \infty$, only the smallest cost c_k will be chosen and freedom of choice attains its minimum $W^{logit}(p) \to -c_k$ where $k = \arg\min_k c_k$. To summarize:

- The advantage is equal to the negative of the composite cost, $W^{logit}(p) = -c + \frac{1}{\beta} H(p) = -\tilde{c}(\beta)$,
- $p_k \to 1/K$, as $\beta \to 0$ and $p_k \to 1$ where $k = \arg\min_k c_k$ as $\beta \to \infty$.
- $c \to \sum_{k=1}^{K} c_k/K$ as $\beta \to 0$ and $c \to c_k$ where $k = \arg\min_k c_k$ as $\beta \to \infty$.
- $H(p) \to \log K$ as $\beta \to 0$ and $H(p) \to 0$ as $\beta \to \infty$.
- $W^{logit}(p) \to \infty$ as $\beta \to 0$ and $W^{logit}(p) \to -c_k$ where $k = \arg\min_k c_k$ as $\beta \to \infty$.

The welfare measure $W^{logit}(p)$ behaves in the limiting cases of the parameter $\beta \to 0$ and $\beta \to \infty$ in accordance with our intuitive understanding of a welfare measure based on freedom of choice and cost.

Similarly, in Sect. 4.5.1 dealing with the classic ARUM approach we noted that as the parameter θ becomes small the expected achieved utility $\tilde{v}$, (4.12) grows without bounds at the same time as the choice probabilities (4.11) become more equal. The growth of the expected achieved utility reflects the fact that the variance of the extreme value distributed random variables X_k, tends to infinity as θ approaches zero. A small value of the parameter θ indicates that the model is poor, since the utilities v_k do not explain much when the choice probabilities become equal. This makes the ARUM interpretation of the composite utility dubious: for small values of θ the logit model is poor but the expected achieved utility, which is dominated by the non observable random variables X_k, is at the same time large.

The expected cost is given by $c = \sum_{k=1}^{K} c_k p_k$. The expected cost is a function of the cost coefficients $[c_k]$ and the probabilities $[p_k]$. The entropy of the probability distribution is given by $H = -\sum_{k=1}^{K} p_k \log p_k$. The entropy H is a property of the probability distribution p. It depends on the costs through the probabilities only. The pair $(-c, H)$ is a two-dimensional benefit measure based on expected negative cost and expected freedom of choice. We have just seen that for the logit model the two-dimensional benefit measure $(-c, H)$ can be condensed into a one-dimensional benefit measure. This combined benefit measure turns out to be the composite utility, well known from classic discrete choice theory where it has the interpretation of expected achieved perceived utility. We used the dimensional properties of the entropy expression to derive the combined benefit measure W. We could equally well have used the fact that for the logit distribution the entropy is a differentiable function of the expected cost c.

6.5.2 Numerical Illustrations

The discussion in Sect. 6.5.1 is illustrated by the following numerical example. The logit probabilities are shown for $\beta = 0$, $\beta = 1$ and $\beta = 2$ in Fig. 6.1. The variation of the value of the entropy $H(\beta)$ as a function of β is given in Fig. 6.2 for

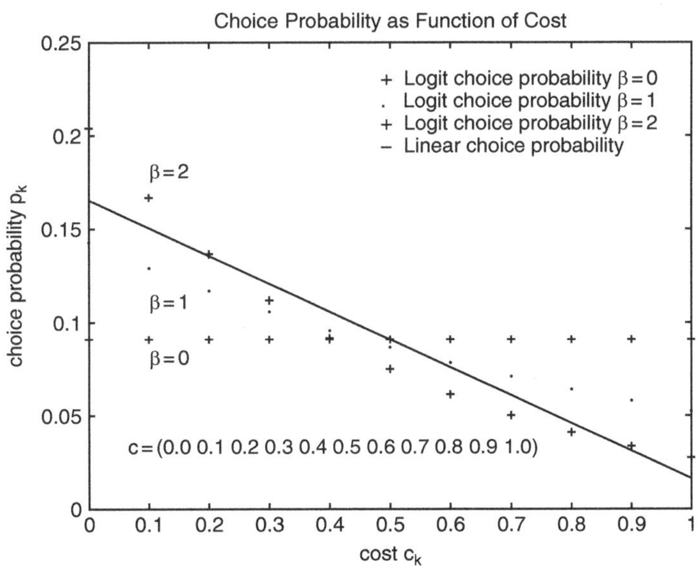

Fig. 6.1 Logit probabilities and one linear choice probability function

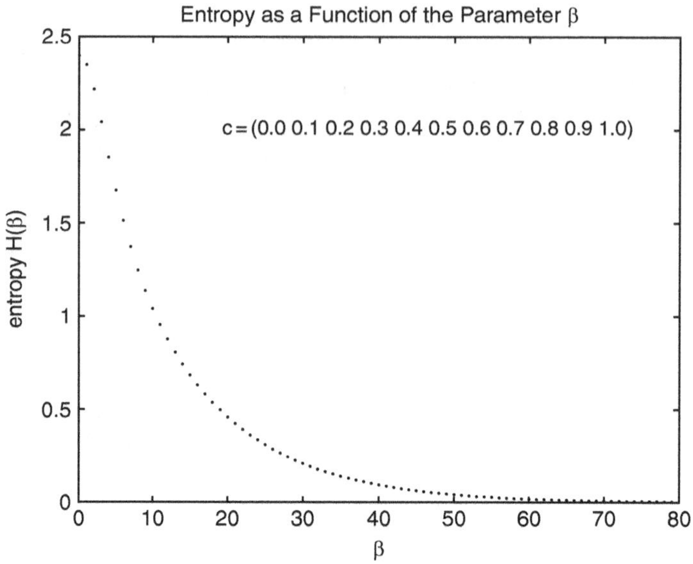

Fig. 6.2 Entropy as a function of the parameter β

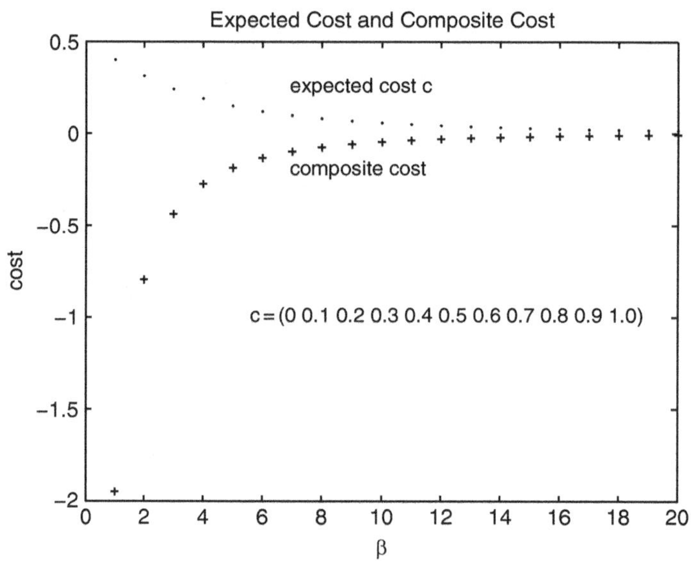

Fig. 6.3 Expected cost $c = \sum_{k=1}^{K} c_k p_k$ and composite cost $\tilde{c}(\beta) = c - \frac{1}{\beta} H(p)$

a numerical example that will be used also in Chapt. 7. Figure 6.3 shows expected cost c and composite cost $\tilde{c}(\beta)$ for different values of β. As the formulas indicates the composite cost is much smaller than the expected cost for small values of the parameter β. On the other hand, the limit of the composite cost tends to the expected cost as $\beta \to \infty$.

6.5.3 Welfare Measure for the General Logit Model

We shall now treat the general logit model (Sect. 4.3). Recall that we have S cost measures and that activity equivalence is defined with respect to M activity measures. The cost matrix $\mathrm{C} \in \mathrm{R}^{S \times K}$ has columns c_k with elements $c_s sk$ denoting the value of the cost function (cost measure) s for alternative k. Similarly, the activity matrix $\mathrm{A} \in \mathrm{R}^{M \times K}$ has has columns a_k with elements a_{mk} indicating the level of activity measure m in alternative k.

The cost-minimizing probability distributions in the general case were given by (5.7) in Proposition 3,

$$p_k = \exp(\mu + \alpha^T a_k - \beta^T c_k),\ k = 1, \ldots, K,$$

which can be written as

$$p_k = \frac{\exp(\alpha^T a_k - \beta^T c_k)}{\sum_{k=1}^K \exp(\alpha^T a_k - \beta^T c_k)},\ k = 1, \ldots, K.$$

Note that $\beta^T c_k$ represents a weighted sum over the cost measures(attributes) for alternative k. Hence, $\beta^T c_k$ can be interpreted as the value of a generalized cost for alternative k.

In order to combine cost and entropy into one measure we have to translate entropy into cost terms. Again, this can be done by a simple dimensional analysis as follows. We obtain

$$\begin{aligned} H(p) &= -\sum_{k=1}^K p_k \log p_k = -\sum_{k=1}^K p_k(\alpha^T a_k - \beta^T c_k) + \log \sum_{k=1}^K \exp(\alpha^T a_k - \beta^T c_k) \\ &= -\alpha^T Ap + \beta^T Cp + \log \sum_{k=1}^K \exp(\alpha^T a_k - \beta^T c_k). \end{aligned}$$

In the general case treated here there are S cost measures or cost attributes. In order to translate entropy into cost we have to select one cost measure as defining the unit of cost of the other cost measures. Let us select the first one. Then by dividing through with the first β value, β_1, we obtain

$$\frac{1}{\beta_1} H(p) = -\frac{1}{\beta_1}\alpha^T Ap + \frac{1}{\beta_1}\beta^T Cp + \frac{1}{\beta_1}\log \sum_{k=1}^K \exp(\alpha^T a_k - \beta^T c_k). \quad (6.9)$$

The second term can be written

$$\frac{1}{\beta_1}\beta^T Cp = \sum_{k=1}^K c_{1k} p_k + \sum_{k=1}^K \sum_{s=2}^S p_k c_{sk} \frac{\beta_s}{\beta_1}.$$

From (6.9) it can be seen that $H(p)/\beta_1$ has the same dimensional property as c_{1k}. Hence entropy divided by the parameter β_1 is measured in the same terms as the first cost measure. But this means that we can combine cost and entropy into one one dimensional measure,

$$W^{general}(p) = -\sum_{k=1}^{K} c_{1k} p_k - \sum_{k=1}^{K}\sum_{s=2}^{S} p_k c_{sk} \frac{\beta_s}{\beta_1} + \frac{H(p)}{\beta_1}$$
$$= -\sum_{k=1}^{K}\sum_{s=1}^{S} p_k c_{sk} \frac{\beta_s}{\beta_1} + \frac{1}{\beta_1} H(p).$$

We give this as a proposition:

Proposition 18. *For the general logit model the combined advantage $W^{general}(p)$ of expected negative generalized cost and expected freedom of choice for the decision makers is equal to*

$$W^{general}(p) = -\sum_{k=1}^{K}\sum_{s=1}^{S} p_k c_{sk} \frac{\beta_s}{\beta_1} + \frac{1}{\beta_1} H(p). \tag{6.10}$$

By combining (6.9) and (6.10) we obtain

$$W^{general}(p) = -\frac{1}{\beta_1}\alpha^T A p + \frac{1}{\beta_1} \log \sum_{k=1}^{K} \exp(\alpha^T a_k - \beta^T c_k). \tag{6.11}$$

In this case composite cost can be defined as follows:

$$\tilde{c}(\alpha, \beta) = -\frac{1}{\beta_1} \log \sum_{k=1}^{K} \exp(\alpha^T a_k - \beta^T c_k).$$

6.5.4 Welfare Measures for some Particular Models

We give the welfare measure for some common models such as the stochastic route choice model, the multi-attribute case, structured (nested) logit models, the gravity model for trip distribution and for models with socioeconomic factors.

6.5.5 Welfare Measure for the Stochastic Route Choice Model

The choice probabilities for the stochastic route choice model with fixed link costs was given in Sect. 5.2,

$$p_k = \frac{\exp(-\beta c_k)}{\sum_{k=1}^{K} \exp(-\beta c_k)}.$$

This is the simple logit model (6.4), hence the combined advantage of expected negative cost and expected freedom of choice is once more given by (6.8),

$$W^{logit}(p) = -c + \frac{1}{\beta}H(p) = -\tilde{c}(\beta).$$

In Sect. 6.5.1 we discussed limiting cases for the logit model. The same results holds for the stochastic route choice model. Hence, for the stochastic route choice model we have as β becomes small $p_k \to 1/K$, as $\beta \to 0$. This means that the traffic load is spread out equally over all routes. The expected cost $c \to \sum_{k=1}^{K} c_k/K$ and the entropy $H(p) \to \log K$ as $\beta \to 0$. And, since the advantage is measured in monetary terms, $W^{logit}(p) \to \infty$ as $\beta \to 0$. If, on the other hand, $\beta \to \infty$, then $p_k \to 1$ where $k = \arg\min_k c_k$. This means that all traffic is concentrated on one route, and $W^{logit}(p) \to -c_k$ where $k = \arg\min_k c_k$ as $\beta \to \infty$. Thus the advantage measure $W^{logit}(p)$ behaves as we would expect.

6.5.6 Welfare Measure for the Multi-Attribute Case

In Sect. 5.3 we showed that in the linear-in-parameters logit model the choice probabilities are given by

$$p_k = \frac{\exp(-\sum_{s=1}^{S} \beta_s c_{ks})}{\sum_{k=1}^{K} \exp(-\sum_{s=1}^{S} \beta_s c_{ks})}, k = 1, \ldots, K, \tag{6.12}$$

where c_{ks}, $k = 1, \ldots, K$, $s = 1, \ldots, S$, represents S costs/attributes, and where β_s, $s = 1, \ldots, S$ are parameters (see). In the absence of activity constraints the measure of welfare/advantage follows from (6.11),

$$W^{multi}(p) = \frac{1}{\beta_1} \log \sum_{k=1}^{K} \exp(\sum_{s=1}^{S} -\beta_s c_{ks}). \tag{6.13}$$

6.5.7 Welfare Measure for Structured (Nested) Logit Models

In Sect. 5.7 we gave two-level nested logit models, in which there is a choice between M alternatives in the upper level, and, conditionally on the choice in the upper level, there is a choice between K alternatives in the lower level. Let the costs be c_{km}, $k = 1, \ldots, K$, and $m = 1, \ldots, M$. We shall here give the welfare measures for the structured logit model with one and several cost constraints, 5.11 and 5.13 respectively. Consider first the structured logit model with one cost constraint:

$$p_{km} = \frac{\exp(\alpha_m - \beta c_{km})}{\sum_{k,m} \exp(\alpha_m - \beta c_{km})}.$$

The entropy for this model becomes

$$H^{structured\ logit}(p) = -\sum_{k,m} \alpha_m p_{km} + \beta \sum_{k,m} c_{km} p_{km} + \log \sum_{k,m} \exp(\alpha_m - \beta c_{km}),$$

and the measure of welfare/advantage becomes

$$\begin{aligned} W^{structured\ logit}(p) &= -\frac{1}{\beta} \sum_{k,m} \alpha_m p_{km} + \frac{1}{\beta} \log \sum_{k,m} \exp(\alpha_m - \beta c_{km}) \\ &= -\frac{1}{\beta} \sum_{k,m} \alpha_m p_{km} - \tilde{c}(\beta), \end{aligned}$$

where

$$\tilde{c}(\beta) = -\frac{1}{\beta} \log \sum_{k,m} \exp(\alpha_m - \beta c_{km}).$$

Similarly, for the structured logit model with several cost constraints:

$$p_{km} = \frac{\exp(\alpha_m - \beta_m c_{km})}{\sum_{k,m} \exp(\alpha_m - \beta_m c_{km})} \quad k = 1, \ldots, K,\ m = 1, \ldots, M.$$

The entropy for this model becomes

$$H^{structured\ logit}(p) = -\sum_{k,m} \alpha_m p_{km} + \sum_{k,m} \beta_m c_{km} p_{km} + \log \sum_{k,m} \exp(\alpha_m - \beta_m c_{km}).$$

Since we have here M possibly different cost measures we have to decide in which cost unit cost shall be measured. Let $m = 1$ define the cost unit. Then total cost is given by $c = \sum_{k=1}^{K} \sum_{m=1}^{M} (\beta_m/\beta_1) c_{km} p_{km}$. The measure of welfare/advantage becomes

$$\begin{aligned} W^{structured\ logit}(p) &= -\frac{1}{\beta_1} \sum_{k,m} \alpha_m p_{km} + \frac{1}{\beta_1} \log \sum_{k,m} \exp(\alpha_m - \beta_m c_{km}) \\ &= -\frac{1}{\beta_1} \sum_{k,m} \alpha_m p_{km} - \tilde{c}(\beta), \end{aligned}$$

where

$$\tilde{c}(\beta) = -\frac{1}{\beta_1} \log \sum_{k,m} \exp(\alpha_m - \beta_m c_{km}).$$

Similarly, for the standard nested logit model (5.16), the probabilities are given by

$$p_{km} = \frac{\exp(\alpha_m - \beta c_{km} - \gamma \tilde{c}_m(\beta))}{\sum_{k,m} \exp(\alpha_m - \beta c_{km} - \gamma \tilde{c}_m(\beta))},$$

and the welfare/advantage measure becomes

$$W^{\textit{standard nested logit}} = -\frac{1}{\beta}\sum_{m=1}^{M}(\alpha_m - \gamma\tilde{c}_m(\beta))\sum_{k=1}^{K} p_{km} + \frac{1}{\beta}\log\sum_{k,m}\exp(\alpha_m - \beta c_{km} - \gamma\tilde{c}_m(\beta)).$$

If β is small, then the parameters α_m dominate. In this case there is little of cost-minimizing behavior.

6.5.8 Welfare Measure for Gravity Model for Trip Distribution

The benefit measure $W(p)$ takes a particularly simple form for the gravity model for trip distribution. Write the gravity model in the form

$$p_{ij} = \exp(\alpha_i + \gamma_j - \beta c_{ij}),$$

where $\sum_{j=1}^{J} p_{ij} = a_i$, and $\sum_{i=1}^{I} p_{ij} = b_j$ and $\sum_{i=1}^{I} a_i = \sum_{j=1}^{J} b_j = 1$.

We obtain

$$H(p) = -\sum_{i=1}^{I}\alpha_i a_i - \sum_{j=1}^{J}\gamma_j b_j + \beta\sum_{i=1}^{I}\sum_{j=1}^{J} c_{ij} p_{ij}$$

and it follows that

$$W^{gravity}(p) = -\frac{1}{\beta}(\sum_{i=1}^{I}\alpha_i a_i + \sum_{j=1}^{J}\gamma_j b_j). \tag{6.14}$$

6.5.9 Welfare Measures for Models with Socioeconomic Factors

In the model with socioeconomic factors given in Sect. 5.9 the choice probability distribution was given by

$$p_{ks} = \frac{\exp(\alpha_s - \beta_s c_{ks})}{\sum_{k=1}^{K}\sum_{s=1}^{S}\exp(\alpha_s - \beta_s c_{ks})}.$$

The entropy becomes

$$H(p) = -\sum_{s=1}^{S} a_s\alpha_s + \sum_{s=1}^{S} \beta_s \sum_{k=1}^{K} c_{ks} p_{ks} + \log \sum_{k=1}^{K} \sum_{s=1}^{S} \exp(\alpha_s - \beta_s c_{ks}),$$

where $a_s = \sum_{k=1}^{K} p_{ks}$.

The advantage becomes

$$W^{socioeconomic} = -\sum_{s=1}^{S} a_s \frac{\alpha_s}{\beta_1} + \frac{1}{\beta_1} \log \sum_{k=1}^{K} \sum_{s=1}^{S} \exp(\alpha_s - \beta_s c_{ks})$$

6.6 Extensions

6.6.1 Extended Benefit Measure

In Proposition 17 the combined benefit of expected negative cost and expected freedom of choice, the advantage, was given for the logit model (6.4). The combined benefit measure W according to (6.7) can be extended to a benefit measure for an arbitrary probability distribution p. This benefit measure has a very interesting property.

Let p be an arbitrary choice probability distribution. Form the extended benefit measure

$$W^{extended}(p) = -c(p) + \frac{1}{\beta} H(p).$$

This is essentially the Lagrangian function for the maximum entropy problem. It follows from Theorem 3 in the Appendix that maximum entropy is obtained for the simple logit model,

$$p_k^m = \frac{\exp(-\beta c_k)}{\sum_{k=1}^{K} \exp(-\beta c_k)}.$$

Hence

$$\begin{aligned} W^{extended}(p) &= -c(p) + \frac{1}{\beta} H(p) \\ \leq max_p[-c(p) + \frac{1}{\beta} H(p)] &= -c(p^p) + \frac{1}{\beta} H(p^m). \end{aligned} \tag{6.15}$$

By using (6.5)

$$-H(p) = \sum_{k=1}^{K} p_k \log p_k = -\beta(c - \tilde{c}(\beta)),$$

we obtain

$$W^{extended}(p) \leq -\tilde{c}(\beta).$$

where $\tilde{c}(\beta)$ is the composite cost. Hence, our extended benefit measure is always less than or equal to the negative of the composite cost $\tilde{c}(\beta)$, and equality holds for the logit model.

- The *extended benefit measure*

$$W^{extended}(p) = -\sum_{k=1}^{K} p_k c_k + \frac{1}{\beta} H(p)$$

  is the weighted sum of the negative of *expected cost* and the *entropy* of the probability distribution (= the expected freedom of choice). It represents the expected benefit, advantage, obtained by a randomly chosen decision maker.
- For any probability distribution p the benefit measure $W^{extended}(p)$ obtained by a randomly chosen decision maker is less than or equal to the composite cost $\tilde{c}(\beta)$, and *equality holds for the logit model.*

Note that the rate of substitution between money and entropy, β, is known only if the probability distribution is the logit model.

In the additive random utility maximizing approach the composite cost has the interpretation of expected achieved perceived cost, except for a constant, for a randomly chosen decision maker, who minimizes his/her perceived cost. By widening the benefit measure to include expected negative cost and expected freedom of choice we obtain the interesting result that the composite cost has the interpretation of a benefit measure including expected utility and expected freedom of choice, irrespective of how the choice probabilities have been derived.

6.6.2 A Lower Bound on Observed Negative Entropy

A lower bound on the observed negative entropy is obtained as follows.

By using the result on the extended benefit measure for an arbitrary probability distribution above, (6.15), we obtain for the observed probability distribution $\bar{p}$

$$-c(\bar{p}) + \frac{1}{\beta} H(\bar{p}) \leq -c(p) + \frac{1}{\beta} H(p),$$

where p is the logit probability. This gives a lower bound on the observed negative entropy

$$-H(p) + \beta c(p) - \beta \bar{c} \leq -H(\bar{p}). \tag{6.16}$$

This will be used in constructing the lower bound in (7.3).

6.6.3 Value of Time

Formula (6.13) above has interesting consequences. In discrete travel choice models the subjective value of time of travel SVTT is calculated as the rate of substitution between time and cost for constant utility (See e.g. Jara-Diaz 2000, p. 309). E.g. if v_{k1} is negative cost $c_k = -v_{k1}$ and v_{k2} is negative time $t_k = -v_{k2}$, then the linear-in -parameters logit model (6.12) becomes

$$p_k = \frac{\exp(-\beta_1 c_k - \beta_2 t_k)}{\sum_{k=1}^{K} \exp(-\beta_1 c_k - \beta_2 t_k)}, k = 1, \ldots, K,$$

and the rate of substitution between time and cost becomes β_2/β_1. With respect to our benefit measure $W(p)$ we obtain the same quotient $\frac{\partial W}{\partial t_k} / \frac{\partial W}{\partial c_k} = \beta_2/\beta_1$.

6.7 Comments

6.7.1 Revealed Freedom of Choice, Diversity of Choice, Flexibility of Choice

We have chosen the term "Freedom of choice" to indicate the richness of the opportunities open to the decision makers. This notion is close to "Diversity of choice" and "Flexibility of choice".

6.7.2 Freedom of choice: Advantage and Achievement

The concept of freedom of choice has been deeply penetrated by Amartya Sen. "The foundational importance of freedom may well be the most far-reaching substantive problem neglected in standard economics" (Sen 1988, p. 294). In particular, Sen discusses the advantage of individual person's choices as going beyond assessment of the achievements. "Advantage has also to take note of the real opportunities faced by the person. Assessment of advantage must, in this view, involve the evaluation of a *set* of potential achievements and not just the actual one."(Sen 1985, p. 33). This raises a number of questions. "Consider a case in which the set from which a person can choose shrinks, but still includes the best element from the larger set. Then, in terms of *achievement*, the person's position might be seen as remaining unaffected (if the person does choose the best in each case), but the freedom enjoyed by the person would have shrunk. It is relevant to ask, in this context, how the value of this 'freedom' may be taken into account" (Sen 1985, p. 43). Sen discusses the possibility to have a two-parameter evaluation consisting of the maximal element of the choice set and the number of elements of the set. However, he finds "the arbitrariness of choosing the *number* of elements as a reflection of the 'extent' of choice makes this a very limited approach" (Sen 1985, p. 44).

Giving a decision maker more choice by adding alternatives might be valuable. According to Arrow there is one serious objection to this approach. "Suppose one adds to the opportunity set an alternative which is clearly unacceptable, perhaps starving to death ..." Arrow (1995) (p. 8). This would not increase the freedom of choice. However, in the following Arrow argues "that an appropriate definition of freedom requires some reference to preferences". This is indeed done in an indirect way by our revealed measure of freedom of choice (6.2) based on the observed entropy because the observed choices are influenced by the cost of the alternatives.

Sen's argument is basically deterministic. Within our probabilistic approach it becomes possible to combine into one single advantage measure achievement in terms of negative cost and freedom of choice in terms of entropy. Sen makes a distinction between two different perspectives when judging a person's position in a social arrangement:

"(1) the actual achievement, and (2) the freedom to achieve" , Sen (1992), p. 31.

See also Sen (1991). We follow Sen in using the distinction between advantage and achievement.

6.7.3 Entropy and Freedom of Choice

Shannon originally introduced axiomatically entropy as a measure of how much "choice" is involved (Shannon 1948, p. 393). The use of entropy as a measure of variation or dispersion has a long history, see e.g. Erlander (1980). Related is the use of entropy as an accessibility measure (Erlander 1977; Schéele 1977). Erlander (1975), studying the trip distribution problem, interprets entropy as the value of freedom of choice (value of choosing an alternative other than the one with highest deterministic utility). Erlander and Schéele (1974) use a closely related probability as "a measure of quality that characterizes the degree of freedom of the travelers' choices" (Erlander and Schéele 1974, p. 589). A thorough discussion was given by Boyce et al. (1988). Suppes (1996), motivated by the property of the entropy to be a measure of variation, proposes the observed entropy of the sample $H([z_k/N])$, (6.2), as the measure of freedom of a set of alternatives and uses this measure to discuss ergodic theory of freedom for discrete stochastic processes. He uses entropy as a measure of the spread or variation of the observed relative frequencies but does not identify, as Erlander (2005), the observed entropy (6.2) as closely related to the numerical representation function the existence of which is guaranteed by Theorem 1 in Suppes (1987). Much of the material on freedom of choice in this chapter was discussed by Erlander (2005). Freedom of choice can be related to variety-seeking behavior (See e.g. Anderson et al. 1992, p. 79).

6.7.4 Welfare and Benefit Measures

Williams (1977) identified the composite cost (6.6) as the consumer surplus function and noted that it can take negative values. This fact was considered by Fisk and Boyce (1984) and others to be inconsistent with its interpretation as expected achieved cost within the ARUM approach. They investigate an alternative expression. This is also done by, e.g., de la Barra (1998). In the approach used in this book negative values of the composite cost (6.6) appear naturally if the expected cost c is larger than the freedom of choice $(1/\beta)H(p)$. The residential location model of Wilson et al. (1981), p. 83, "includes a benefit contribution due to the distribution of utility values." In fact, Wilson et al. (1981), p. 56, include entropy in a generalized surplus which is equivalent to (6.7) and interpret entropy as consumer surplus from travel (Wilson et al. 1981, p.168). Erlander and Stewart (1990), p. 148–150, discuss (6.14) as a benefit measure. For the logit model the observed combined advantage $W^{logit}(\bar{p})$ of negative average cost and revealed freedom of choice for the decision makers, (6.8), was given by Erlander (2005). Miyagi and Morisugi (1996) showed, by using the additive random utility maximizing approach, that $\sum_{k=1}^{K} p_k v_k + \frac{1}{\beta} H(p) = \tilde{v}(\beta)$, a fact that is implicit in much of the random utility literature. They interpret entropy as the freedom of choice. A different approach is used by Mattsson and Weibull (2002). They assume that in addition to the expected payoff there is a dis utility or control cost. The latter is formalized axiomatically leading to an entropy expression. The decision maker is assumed to maximize the sum of expected payoff and weighted entropy. This is the same as maximizing our benefit measure $W(p) = v + (1/\beta)H(p)$, resulting in the logit formula (4.11). Maximizing the benefit $W(p)$ is in fact in principle equivalent to the optimization formulations used e.g. by Boyce (1998) in deriving transportation planning models with entropy expressions in the objective or as constraints.

6.8 Notes

This chapter is based on Erlander (2005).

Chapter 7
Graphical Tests of Cost-Minimizing Behavior in Logit Models

Abstract Graphical tests are given for testing the agreement of basic assumptions about cost-minimizing behavior against observations. Simulation is used to study the properties of the graphical test for the simple (multinomial) logit model.

7.1 Introduction

The classical likelihood ratio test is one way of investigating if a probability distribution $p = (p_1, \ldots, p_K)$ is compatible with experience. Let z_k be the number of times decision k is taken in an independent sample of size N from the probability distribution p. Let z denote the vector $(z_1, \ldots, z_K)$, $z \in Z_+^K$. The maximum likelihood estimate of the probability distribution p is equal to the relative frequencies $z_k/N = \bar{p}_k$, $k = 1, \ldots, K$ if nothing more is assumed about the form of p. The log-likelihood ratio test statistic becomes $w(p) = 2[\sum_{k=1}^K z_k \log \bar{p}_k - \sum_{k=1}^K z_k \log p_k]$. This can be used for testing any of the choice probability distributions given in Chaps. 4 and 5. The likelihood ratio test investigates the goodness of a given probability distribution p irrespective of how the probabilities have been derived.

However, in this book we derive the choice probability functions from basic underlying assumptions regarding the behavior of the decision makers. The form of the choice probability function is derived from cost-minimizing behavior. Thus it is of interest to investigate these underlying assumptions about behavior, and not only the derived choice probabilities. We shall now show how the underlying behavioral assumptions about cost-minimizing behavior can be tested.

S. Erlander, *Cost-Minimizing Choice Behavior in Transportation Planning*, Advances in Spatial Science, DOI 10.1007/978-3-642-11911-8_7,

7.2 Testing for Cost-Minimizing Behavior in the Simple Logit Model

The choice probabilities for the logit model was obtained in (4.4) as

$$p_k = \frac{\exp(-\beta c_k)}{\sum_{k=1}^{K} \exp(-\beta c_k)}.$$

The definition of cost-minimizing behavior, Definition 1, can be reformulated as follows:

Definition 18 (Cost-Minimizing Behavior). The probability distribution p represents cost-minimizing behavior if and only if the likelihood of observing z decreases (does not increase) as the average cost of the sample increases (does not decrease), hence the likelihood $\mathcal{L}(z) = \prod_{k=1}^{K} p_k^{z_k}$ decreases as $\bar{c} = \frac{1}{N}\sum_{k=1}^{K} z_k c_k$ increases.

This way of formulating the basic underlying assumption about cost-minimizing behavior can be used to construct a graphical test for cost-minimizing behavior. Replacing the unknown probability distribution p with its maximum likelihood estimate $\bar{p} = (z_1/N, \ldots, z_K/N)$, if nothing more is assumed about the probability distribution, the log likelihood can be written

$$\sum_{k=1}^{K} z_k \log(z_k/N) = -NH(\bar{p}).$$

Let there be J independent samples $z_k^j, k = 1, \ldots, K,\ j = 1, \ldots, J$ of size N. (If enough numbers of observed samples is not available, resampling technique can be used to generate the J samples from one original sample.) For each sample j we obtain the pair of observations $(\bar{c}^j, H(\bar{p}^j))$.

- If cost-minimizing behavior holds, then a graphical plot of the negative observed sample entropies $-H(\bar{p}^j)$ against the observed average sample costs $\bar{c}^j$ would show a pattern of decreased negative entropy for increased sample cost.

Hence, a graphical test can be constructed by plotting the observed negative entropy $-H(\bar{p}^j)$ against the observed average cost $\bar{c}^j$; increasing values of average cost $\bar{c}^j$ implies a pattern of decreasing values of the observed negative entropy $-H(\bar{p}^j)$. For the series of samples $j = 1, \ldots, J$, the observed negative entropy $-H(\bar{p}^j)$ will appear as a stochastic variable taking values related in a probabilistic way to the average cost values $\bar{c}^j$, which varies in a probabilistic way as well. Hence, we need to know how the pattern looks at least for large samples. However, our interest lies in the random variation of the observed negative entropy, not in the random variation of the average cost values. Hence, we shall study the variation of the observed negative entropy conditionally on the value of the average cost. This can be done since cost-minimizing behavior is defined for any value of the average cost. This is equivalent of assuming that the average cost is fixed at the observed value for each sample in the proposition below.

If cost-minimizing behavior is at hand, then, according to Definition 18, the likelihood function is decreasing in the observed average cost. We shall use this property to construct a graphical test of the statement in Definition 18.

Let there be J independent samples $z_k^j, k = 1, \ldots, K,\ j = 1, \ldots, J$ of size N. The maximum likelihood estimate of the probability p_k in sample j is the relative frequency $\bar{p}_k^j = z_k^j/N, k = 1, \ldots, K$. Let $\bar{p}^j = (\bar{p}_1^j, \ldots, \bar{p}_K^j)$. Let the entropy of the probability distribution p be defined by $H(p) = -\sum_{k=1}^K p_k \log p_k$, and correspondingly for the observed distribution $\bar{p}^j$, $H(\bar{p}^j) = -\sum_{k=1}^K \bar{p}_k^j \log \bar{p}_k^j$. Let $c = \sum_{k=1}^K p_k c_k$, and $\bar{c}^j = \sum_{k=1}^K \bar{p}_k^j c_k$. For each sample j we obtain the pair of observations $(\bar{c}^j, H(\bar{p}^j))$.

From Proposition 19 below we get an approximate expression for the random variations for each sample j,

$$-H(\bar{p}^j) \approx -H(p) + \beta c - \beta \bar{c}^j + \frac{1}{2N}\chi^2,$$

where the stochastic variable χ^2 has a χ^2-distribution with $(K-2)$ degrees of freedom. We shall use this expression in order to construct approximate upper and lower bounds for the random variation conditionally on $\bar{c}^j$. We have to estimate the unknown true values of of the probability distribution p, the expected cost c and the parameter value β. If the number J of samples is not too small, we may estimate these values from the total sample. Neglecting the dependence between the values obtained for the total sample and the corresponding values for the sub samples, we replace the true but unknown values of the probability p with the observed relative frequencies $\bar{p}_k = \frac{1}{JN}\sum_{j=1}^J z_k^j$ for the total sample, the expected unknown cost c with the average observed cost $\bar{c} = \frac{1}{JN}\sum_{j=1}^J \sum_{k=1}^K c_k z_k^j$ for the total sample and the likewise unknown parameter value β with its maximum likelihood estimate $\hat{\beta}$, for the total sample, obtained as the solution to $\bar{c} = c$ according to (9.5), obtaining

$$-H(\bar{p}^j) \approx -H(\bar{p}) + \hat{\beta}\bar{c} - \hat{\beta}\bar{c}^j + \frac{1}{2N}\chi^2. \tag{7.1}$$

This will be used to obtain approximative upper and lower bounds for the random variation of $(-H(\bar{p}^j))$. The upper bound is obtained by selecting the appropriate upper tail value of the χ^2-distribution. The lower bound is obtained from 6.16. Hence we obtain the approximative bounds

$$-H(\bar{p}) + \hat{\beta}\bar{c} - \hat{\beta}\bar{c}^j \leq -H(\bar{p}^j) \leq -H(\bar{p}) + \hat{\beta}\bar{c} - \hat{\beta}\bar{c}^j + \frac{1}{2N}\chi^2_\alpha. \tag{7.2}$$

A graphical plot of $-H(\bar{p}^j)$ against $\bar{c}^j$ will cluster around a straight line with slope $-\hat{\beta}$ and intercept $-H(\bar{p}) + \hat{\beta}\bar{c}$, and the stochastic deviations from the line will be distributed as $(1/2N)\chi^2(K-2)$. If the cost-minimizing behavior according to Definition (18) does not hold, the points will not cluster around a line or the slope of the line will be zero. By using (7.2) we obtain the graphical test below, since

$$\Pr[-H(\bar{p})-\hat{\beta}(\bar{c}^j-\bar{c}) \leq -H(\bar{p}^j) \leq -H(\bar{p})-\hat{\beta}(\bar{c}^j-\bar{c})+\chi^2_\alpha(K-2)/2N] \approx (1-\alpha).$$

7.2.1 Graphical Test of Cost-Minimizing Behavior

- For $j = 1, \ldots, J$, plot the observed negative sample entropies $[-H(\bar{p}^j)]$ against the average sample cost $\bar{c}^j$.
- If cost-minimizing behavior according to Definition 18 holds, then for large N, the expected number of points falling between the bounds defined by

$$-H(\bar{p}) + \hat{\beta}\bar{c} - \hat{\beta}\bar{c}^j \leq -H(\bar{p}^j) \leq -H(\bar{p}) + \hat{\beta}\bar{c} - \hat{\beta}\bar{c}^j + \frac{1}{2N}\chi^2_\alpha \tag{7.3}$$

  is approximately $(1-\alpha)J$. Here $\chi^2_\alpha = \chi^2_\alpha(K-2)$ is the upper α-tail of a χ^2-distribution with $(K-2)$ degrees of freedom.

We accept the hypotheses that cost-minimizing behavior holds if $(1-\alpha)J$ points fall within the bounds. If more than αJ points fall outside of the bounds, the hypotheses is rejected.

Experiments with simulated data are shown in the attached figures in Sect. 7.2.4.

7.2.2 Asymptotic Distribution of the Observed Negative Entropy

We shall drop the j-indices since the proposition below will hold for each j. We shall prove the proposition below. Let the expected cost $c = \sum_{k=1}^{K} c_k p_k$ and the average cost $\bar{c} = \frac{1}{N}\sum_{k=1}^{K} z_k c_k$.

Proposition 19 (Asymptotic Distribution. Logit Model Case). *Assume that cost-minimizing behavior according to Definition 18 (or equivalently, assumption (4.3)) is true. Then*

$$p_k = \frac{\exp(-\beta c_k)}{\sum_{k=1}^{K}\exp(-\beta c_k)}.$$

Furthermore, for large N, conditionally on the average cost $\bar{c}$, the observed negative entropy can be written approximately as

$$-H(\bar{p}) \approx -H(p) + \beta c - \beta\bar{c} + \frac{1}{2N}\chi^2,$$

where the stochastic variable χ^2 has a χ^2-distribution with $(K-2)$ degrees of freedom. The expected value and the variance of the observed negative entropy are approximately given by

$$E[-H(\bar{p})] \approx -H(p) + \beta c - \beta\bar{c} + \frac{K-2}{2N} = \beta\tilde{c}(\beta) - \beta\bar{c} + \frac{K-2}{2N}, \textit{ and}$$

$$Var[-H(\bar{p})] \approx \frac{K-2}{2N^2}.$$

Proof. The first statement was proved in Proposition 1. The rest of the proposition follows in the same way as in Sect. 6.4. We obtain

$$-H(\bar{p}) = \sum_{k=1}^{K} (z_k/N)\log(z_k/N) = \sum_{k=1}^{K} (z_k/N)\log p_k (z_k/Np_k)$$

$$= \sum_{k=1}^{K} (z_k/N)\log p_k + \frac{1}{N}\sum_{k=1}^{K} z_k \log(z_k/Np_k).$$

For the logit model (4.4) the first term can be written

$$\sum_{k=1}^{K} (z_k/N)\log p_k = -\beta\sum_{k=1}^{K} (z_k/N)c_k - \log\sum_{k=1}^{K} \exp(-\beta c_k) = -\beta\bar{c} + \beta c - H(p),$$

where we have used (6.5),

$$-\log\sum_{k=1}^{K} \exp(-\beta c_k) = \beta c - H(p).$$

The second term can be approximated by

$$\frac{1}{N}\sum_{k=1}^{K} z_k \log(z_k/Np_k) \approx \frac{1}{2N}\sum_{k=1}^{K} \frac{(z_k - Np_k)^2}{Np_k}, \tag{7.4}$$

and we obtain,

$$-H(\bar{p}) \approx -H(p) + \beta c - \beta\bar{c} + \frac{1}{2N}\sum_{k=1}^{K} \frac{(z_k - Np_k)^2}{Np_k}. \tag{7.5}$$

The observed frequencies (z_k) have a multinomial probability distribution, and hence, for large N, z_k is approximately normal with expected value $= Np_k$ and variance $= Np_k(1-p_k) \approx Np_k$. Hence, for large N,

$$\sum_{k=1}^{K} \frac{(z_k - Np_k)^2}{Np_k}$$

has a χ^2-distribution with $(K - 2)$ degrees of freedom since there are two linear relations: $\sum_{k=1}^{K} z_k = N$ and $\bar{c} = \frac{1}{N} \sum_{k=1}^{K} z_k c_k$. Writing

$$\sum_{k=1}^{K} \frac{(z_k - Np_k)^2}{Np_k} = \chi^2,$$

we obtain from (7.5) for the observed negative entropy

$$-H(\bar{p}) \approx -H(p) + \beta c - \beta \bar{c} + \frac{1}{2N} \chi^2.$$

The rest of the proposition follows since the expected value and variance of a χ^2-distributed variable with K degrees of freedom are, respectively, $E[\chi^2(K)] = K$ and $Var[\chi^2(K)] = 2K$. □

7.2.3 Simulation Experiments

Simulation studies have been performed in the following way for the logit model.

For different values of the parameter β, $(\beta = 0, 1, 2)$, $N = 500$ observations have been generated from the logit probability distribution

$$p_k = \exp(-\beta c_k) / \sum_{k=1}^{K} \exp(-\beta c_k), k = 1, \ldots, K,$$

with cost vector $(0.0, 0.1, 0.2, 0.3, 0.4, 0.5, 0.6, 0.7, 0.8, 0.9, 1.0)$. The case $\beta = 0$ is a degenerated logit. Figure 6.1 shows the logit probabilities as well as one linearly decreasing probability function that has been used in the simulations.

The observed negative entropy $-H(\bar{p}^j) = \sum_{k=1}^{K} (z_k^j / N) \log(z_k^j / N)$ has been plotted against the observed average cost $\bar{c}^j = \sum_{k=1}^{K} (z_k^j / N) c_k$.

This has been repeated for $J = 100$ times.

Confidence bounds according to (7.3) have been calculated with the confidence level $= 0.95$. 5 % of the points should on the average fall outside of the bounds. Hence, with $J = 100$ we would expect about five simulated points to fall outside of the bounds for each case.

Two non logit models have also been simulated. The first with permuted cost values and the second with a linearly decreasing probability function.

7.2.4 Figures

Figure 7.1 shows the result for the simulated logit model case. There is a clear linear structure. The number of points falling outside the confidence bounds is close to what was to be expected. Thus the graphical test would in this case conclude that cost-minimizing behavior holds, which, indeed, is true.

In Fig. 7.2 the same plot has been done for the case where the probability distribution is not logit. The probabilities are the same as in Fig. 7.1 but the probabilities have been assigned to a permutation $w = (0.2, 0.1, 0.7, 1.0, 0.0, 0.9, 0.3, 0.4, 0.6, 0.5, 0.8)$ of the costs. In this way the entropy of the distribution is preserved, whereas the expected cost will change. In this case $\bar{c}^j$ in (7.3) will be replaced by $\bar{w}^j = \sum_{k=1}^{K} \bar{p}_k^j w_k$. About two third of the points fall outside the confidence bounds for $\beta = 1$ and $\beta = 2$. Thus the graphical test gives strong evidence that the cost-minimizing hypothesis does not hold, which is also true. Still there is some linear structure in the observations.

Another non logit case is shown in Fig. 7.3. Here the probabilities decrease linearly as is shown in Fig. 6.1. The number of points falling outside of the confidence bounds is considerably larger for the linearly decreasing probabilities than for the logit model. From the figure it is clear that even this slight deviation from the logit model shows up in the graphical plot.

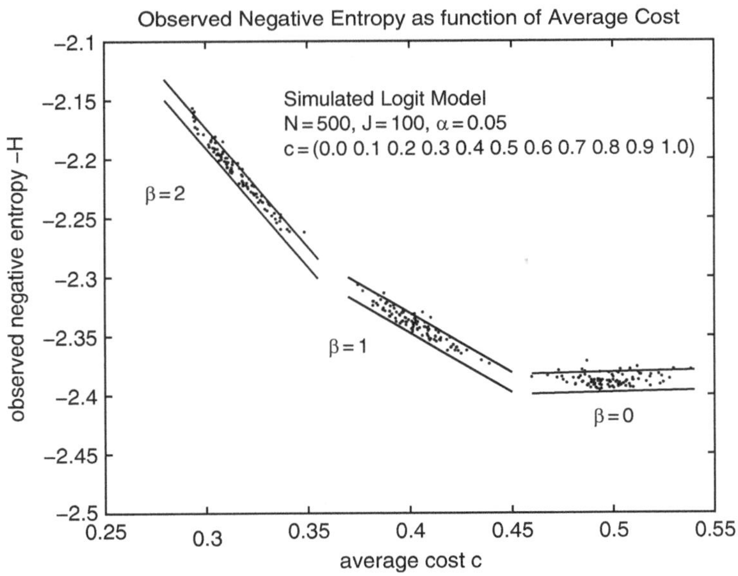

Fig. 7.1 Simulated logit model. $p_k = \frac{\exp(-\beta c_k)}{\sum_{k=1}^{K} \exp(-\beta c_k)}, k = 1, \ldots, K$

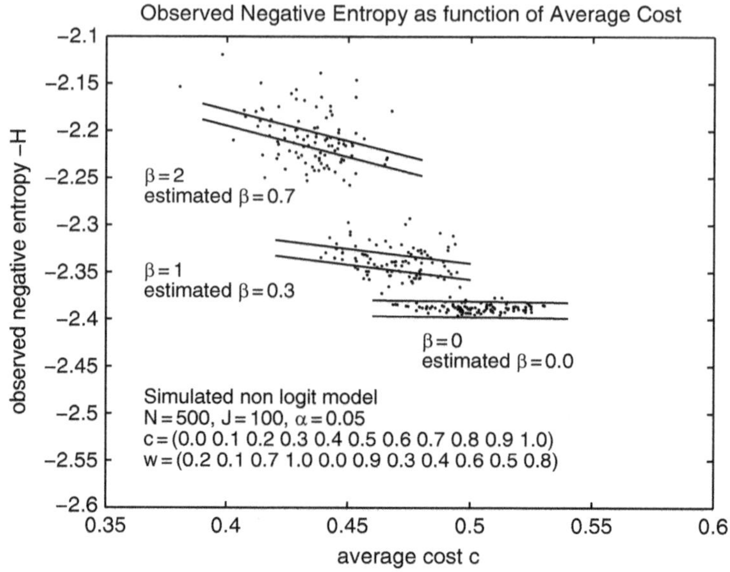

Fig. 7.2 Simulated non logit model. $p_k = \frac{\exp(-\beta w_k)}{\sum_{k=1}^{K} \exp(-\beta w_k)}, k = 1, \ldots, k$

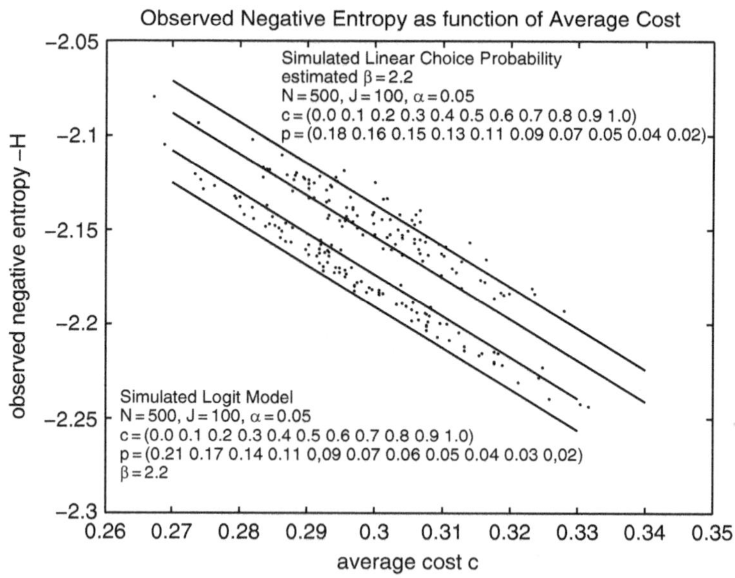

Fig. 7.3 Comparison between logit and linear choice probability

7.2.5 Final Remarks

From the figures we may conclude that, for the cases treated here, the negative of the observed entropy is closely linear in the observed average cost, thus indicating that cost-minimizing behavior is true, if the logit model holds. If the logit model is not true, then, as expected, the stochastic variation is much larger, and cost-minimizing behavior does not hold.

We have shown how the suggested graphical test may be used to test if cost-minimizing behavior holds, and thus if the logit model can be justified by this hypothesis. However, the logit model (4.4) implies cost-minimizing behavior according to Definition 18. Hence, *if an application of our graphical test rejects cost-minimizing behavior, then we have to reject the logit model as well.* This means that the graphical test can be used as a test of the logit model itself irrespective of its derivation.

7.3 Multi-Attribute Discrete Choice Models

In Sect. 7.2 we showed how a graphical test for the simple logit model can be constructed. We shall here construct the corresponding graphical test for the multi-attribute model.

In Sect. 5.3 the multi attribute logit model was given as

$$p_k = \frac{\exp(-\sum_{s=1}^{S}\beta_s c_{sk})}{\sum_{k=1}^{k}\exp(-\sum_{s=1}^{S}\beta_s c_{sk})}, \quad k = 1, \ldots, k, \tag{7.6}$$

where $\beta_s \geq 0, \quad s = 1, \ldots, s$. Let there be n independent observations $\mathbf{z} = (z_1, \ldots, z_k)$. The expected generalized cost c was defined in (5.3) by

$$c = \sum_{k=1}^{K}\sum_{s=1}^{S}\beta_s c_{sk} p_k,$$

and the average generalized cost $\bar{c}$ is defined correspondingly

$$\bar{c} = \sum_{k=1}^{K}\sum_{s=1}^{S}\beta_s c_{sk}(z_k/N).$$

Similarly, let

$$c_s = \sum_{k=1}^{k} c_{sk} p_k \text{ and } \bar{c}_s = \sum_{k=1}^{k} c_{sk}(z_k/N).$$

The results for the simple logit model can be extended as follows. let $\bar{p}_k = z_k/n$ and $\bar{p} = (\bar{p}_1, \ldots, \bar{p}_K)$. We have for the entropy and the observed entropy, respectively

$$H(p) = -\sum_{k=1}^{K} p_k log p_k$$

and

$$H(\bar{p}) = -\sum_{k=1}^{K} \bar{p}_k log \bar{p}_k = \sum_{k=1}^{K} (z_k/N) log(z_k/N).$$

In the same way as in Proposition 19 we obtain the asymptotic distribution:

Proposition 20 (Asymptotic Distribution). *Let there be N independent observations $\mathbf{z} = (z_1, \ldots, z_K)$. Assume that cost-minimizing behavior holds according to Definition 18. Then, for large N, conditionally on the average costs $\bar{c}_s$, $s = 1, \ldots, S$, the observed negative entropy can be written approximately as*

$$-H(\bar{p}) \approx -H(p) + \sum_{s=1}^{S} \beta_s c_s - \sum_{s=1}^{S} \beta_s \bar{c}_s + \frac{1}{2N} \chi^2,$$

where the stochastic variable χ^2 has a χ^2-distribution with $(K - 1 - S)$ degrees of freedom. By using the generalized cost this can be written

$$-H(\bar{p}) \approx -H(p) + c - \bar{c} + \frac{1}{2N} \chi^2.$$

The number of degrees of freedom has to be reduced by $(1 + S)$, since there are the $(1 + S)$ constraints $\sum_{k=1}^{K} z_k = N$ and $\bar{c}_s = \sum_{k=1}^{K} c_{sk}(z_k/N)$, s = 1, ..., S.

In the same way as in Sect. 7.2 we obtain a graphical test.

7.3.1 Graphical Test of Cost-Minimizing Behavior

Let there be J independent samples z_k^j, $k = 1, \ldots, K$, $j = 1, \ldots, J$ of size N. The maximum likelihood estimate of the probability p_k in sample j is the relative frequency $\bar{p}_k^j = z_k^j/N$, $k = 1, \ldots, K$. Let $\bar{p}^j = (\bar{p}_1^j, \ldots, \bar{p}_K^j)$. Let the entropy of the probability distribution p be defined by $H(p) = -\sum_{k=1}^{K} p_k \log p_k$, and correspondingly for the observed distribution $\bar{p}^j$, $H(\bar{p}^j) = -\sum_{k=1}^{K} \bar{p}_k^j \log \bar{p}_k^j$. Let $c = \sum_{k=1}^{K} p_k c_k$, and $\bar{c}^j = \sum_{k=1}^{K} \bar{p}_k^j c_k$. For each sample j we obtain the pair of observations $(\bar{c}^j, H(\bar{p}^j))$.

The parameter values $\hat{\beta}_m$ are estimated by solving the maximum likelihood equations (9.5),

$$\sum_{k=1}^{K} p_k = \frac{1}{J} \sum_{j=1}^{J} \sum_{k=1}^{K} (z_k^j / N) = 1,$$

and

$$\sum_{k=1}^{K} p_k c_{ks} = \frac{1}{J} \sum_{j=1}^{J} \sum_{k=1}^{K} (z_{ks}^j / N) c_{ks}, \; s = 1, \ldots, S.$$

The probability p_k is estimated by $\hat{p}_k$ obtained from (7.6) by inserting $\hat{\beta}_s$ for $s = 1, \ldots, S$, and the costs $\hat{c}_s = \sum_{k=1}^{K} \hat{p}_k c_{sk}$ and $\hat{c} = \sum_{s=1}^{S} \hat{\beta} \hat{c}_s$.

- For $j = 1, \ldots, J$, plot the observed negative sample entropies $[-H(\bar{p}^j)]$ against the average sample cost $\bar{c}_s^j$ for each $s = 1, \ldots, S$.
- The projection of the lower and upper bounds on each s-plane for $s = 1, \ldots, S$ gives

$$-H(\bar{p}) + \sum_{s=1}^{S} \hat{\beta}_s c_s - \sum_{r \neq s}^{S} \hat{\beta}_r \bar{c}_r^j - \hat{\beta}_s \bar{c}_s^j \leq$$

$$-H(\bar{p}^j) \leq -H(\bar{p}) + \sum_{s=1}^{S} \hat{\beta}_s c_s - \sum_{r \neq s}^{S} \hat{\beta}_r \bar{c}_r^j - \hat{\beta}_s \bar{c}_s^j + \frac{1}{2N} \chi_\alpha^2.$$

- If cost-minimizing behavior according to Definition 18 holds, then for large N, the expected number of points falling between the bounds defined by

$$-H(\bar{p}) + \hat{c} - \bar{c}^j \leq -H(\bar{p}^j) \leq -H(\bar{p}) + \hat{c} - \bar{c}^j + \frac{1}{2N} \chi^2,$$

  is approximately $(1 - \alpha)J$. Here $\chi_\alpha^2(K - 1 - S)$ is the upper α-tail of a χ^2-distribution with $(K - 1 - S)$ degrees of freedom.

7.4 Structured Logit Models

In Sect. 5.7.1 we presented some structured logit models derived through the new definition of cost-minimizing behavior and also showed how the standard nested logit can be derived in this approach. Here we shall give a graphical test for the joint logit model (5.11),

$$p_{km} = \frac{\exp(\alpha_m - \beta c_{km})}{\sum_{k,m} \exp(\alpha_m - \beta c_{km})} \quad k = 1, \ldots, K, \; m = 1, \ldots, M.$$

Let z_m and z_{km} be the number of times alternative m and km are chosen in the upper and lower levels, respectively. Clearly, $z_m = \sum_{k=1}^{K} z_{km}$ and $N = \sum_{m=1}^{M} z_m = \sum_{k=1}^{K} \sum_{m=1}^{M} z_{km}$. Let $p = [p_{km}]$ be the choice probability distribution and $\bar{p} = [z_{km}/N]$ the observed relative frequencies.

For each $m = 1, \ldots, M$ define expected cost and average cost respectively by

$$c_m = \sum_{k=1}^{K} c_{km} p_{km} \quad \text{and} \quad \bar{c}_m = \sum_{k=1}^{K} c_{km}(z_{km}/N).$$

Expected cost and average cost for the whole sample are given by

$$c = \sum_{k=1}^{K} \sum_{m=1}^{M} c_{km} p_{km} \quad \text{and} \quad \bar{c} = \sum_{k=1}^{K} \sum_{m=1}^{M} c_{km}(z_{km}/N).$$

Similarly,

$$\alpha = \sum_{k=1}^{K} \sum_{m=1}^{M} p_{km}\alpha_m \quad \text{and} \quad \bar{\alpha} = \sum_{k=1}^{K} \sum_{m=1}^{M} z_{km}\alpha_m.$$

Entropy and observed entropy are given by

$$H(p) = -\sum_{k=1}^{K} \sum_{m=1}^{M} p_{km} \log p_{km} \quad \text{and} \quad H(\bar{p}) = -\sum_{k=1}^{K} \sum_{m=1}^{M} (z_{km}/N) \log(z_{km}/N).$$

Cost-minimizing behavior was given in Definition 12 for the joint logit model in the following way.

A probability distribution $p = (p_{km}, k = 1, \ldots, K,\ m = 1, \ldots, M)$ represents cost minimizing behavior if and only if, for any sample size N,

$$\left.\begin{array}{l} \sum_{k=1}^{K} z_{km}^1 = \sum_{k=1}^{K} z_{km}^2,\ m = 1, \ldots, M, \text{ and} \\ \sum_{k=1}^{K} \sum_{m=1}^{M} c_{km} z_{km}^1 \le \sum_{k=1}^{K} \sum_{m=1}^{M} c_{km} z_{km}^2, \end{array}\right\} \Longrightarrow \prod_{k=1}^{K} \prod_{m=1}^{M} p_{km}^{z_{km}^1} \ge \prod_{k=1}^{K} \prod_{m=1}^{M} p_{km}^{z_{km}^2}.$$

As before we need the asymptotic distribution of the observed entropy $H(\bar{p})$.

Proposition 21 (Asymptotic Distribution). *Let there be N independent observations $\mathbf{z} = (z_1, \ldots, z_K)$. Assume that cost-minimizing behavior holds according to Definition 12. Then, for large N, conditionally on the average values $\bar{c}$ and $\bar{\alpha}_m$, $m=1, \ldots, M$, the observed negative entropy can be written approximately as*

$$-H(\bar{p}) \approx -H(p) + \beta c - \beta \bar{c} - \sum_{k=1}^{K} \sum_{m=1}^{M} \alpha_m p_{km} + \sum_{k=1}^{K} \sum_{m=1}^{M} \alpha_m (z_{km}/N) + \frac{1}{2N} \chi^2,$$

which can be written

$$-H(\bar{p}) \approx -H(p) + \beta c - \beta \bar{c} - \sum_{m=1}^{M} \alpha_m + \sum_{m=1}^{M} \bar{\alpha}_m + \frac{1}{2N} \chi^2,$$

where the stochastic variable χ^2 has a χ^2-distribution with $(KM - 1 - M)$ degrees of freedom.

7.4.1 Graphical Test of Cost-Minimizing Behavior

We are now ready to give a graphical test for the cost-minimizing assumption for the joint logit model.

Let there be J independent samples (z_{km}^j, $k = 1, \dots, K$, $m = 1, \dots, M$) of size N. For each sample j we obtain the pair of observations $(\bar{c}^j, H(\bar{p}^j))$.

The parameter values $\hat{\alpha}_m$ and $\hat{\beta}$ are estimated by solving the maximum likelihood equations (9.5),

$$\sum_{k=1}^{K}\sum_{m=1}^{M} p_{km} = \frac{1}{J}\sum_{j=1}^{J}\sum_{k=1}^{K}\sum_{m=1}^{M}(z_{km}^j/N) = 1,$$

$$\sum_{k=1}^{K} p_{km} = \frac{1}{J}\sum_{j=1}^{J}\sum_{k=1}^{K}(z_{km}^j/N), \quad m = 1, \dots, M,$$

and

$$\sum_{k=1}^{K}\sum_{m=1}^{M} p_{km}c_{km} = \frac{1}{J}\sum_{j=1}^{J}\sum_{k=1}^{K}\sum_{m=1}^{M}(z_{km}^j/N)c_{km}.$$

- For $j = 1, \dots, J$, plot the observed negative sample entropies $[-H(\bar{p}^j)]$ against the average sample cost $\bar{c}^j$.
- If cost-minimizing behavior according to Definition 12 holds, then for large N, the expected number of points falling between the bounds defined by

$$-H(\bar{p}) - \hat{\beta}(\hat{c} - \bar{c}^j) + \sum_{k=1}^{K}\sum_{m=1}^{M}\hat{\alpha}_m(\hat{p}_{km} - (z_{km}/N)) \le -H(\bar{p}^j)$$

$$\le -H(\bar{p}) - \hat{\beta}(\hat{c} - \bar{c}^j) + \sum_{k=1}^{K}\sum_{m=1}^{M}\hat{\alpha}_m(\hat{p}_{km} - (z_{km}/N)) + \frac{1}{2N}\chi^2,$$

  is approximately $(1-\alpha)J$. Here χ^2 is the upper α-tail of a χ^2-distribution with $(KM - 1 - M)$ degrees of freedom.

7.5 Comments and Extensions

The graphical tests were based on J independent samples $z_k^j, k = 1, \dots, K,\ j = 1, \dots, J$ of size N. The same constructions can be used by resampling from one original data set. The basic idea behind resampling is that the observed frequencies z_k/N will approach the true probabilities p_k as N increases. Hence, resampling from the observed frequency distribution will be close to sampling from the true probabilities if the number of observations N is large. (See e.g. Hjorth 1994.)

Tests for logit-type probabilities in the literature are based on the derived logit probabilities (4.4) and do not go into the underlying models, as given e.g. in Sect. 4.5.1 for the ARUM approach. The reason for this is clearly that the underlying models are formulated in terms of non observable random variables. If data refute the logit probabilities (4.4), then the underlying model used in the derivation of this formula has to be refuted as well. On the other hand, if data does not refute the logit formula, this does not prove that the underlying detailed model is correct; it only shows that the ARUM derivation is one of several candidates for how to formulate the underlying assumptions.

The situation is different when using the approach used in this book. Here cost-minimizing behavior can be studied directly since agreement or disagreement with the definition of cost-minimizing behavior according to (4.3) shows up in the graphical test given in this Chapter. Note that if the graphical test refutes cost-minimizing behavior, then the logit formula (4.4) has to be refuted as well because of Proposition 1.

7.6 Notes

Many methods have been used for testing logit type probabilities. These approaches include using the likelihood ratio test for testing the logit probabilities (4.4), testing for alternatives formulated as GEV models and nested models (Hausman and McFadden 1984; Small and Hsiao 1985; McFadden 1987; Small 1994). Another approach has been testing for independence of irrelevant alternatives (e.g. McFadden et al. 1977, 1981; Fry and Harris 1998).

For (7.4) see e.g. Kullback (1959), p. 113–114.

Acknowledgements P-O Lindberg, L-G Mattsson, C Persson, E Ruist and T E Smith gave valuable comments on earlier versions of this chapter. U Hjorth wrote the first version of the MATLAB random sampling generator used in the simulations.

Part II
Equilibrium

We shall in Part II be concerned with the choice probability formulation in a network with volume dependent link costs.

Chapter 8
Equilibrium

Abstract The definition of cost-minimizing behavior is used to study equilibrium with flow dependent separable link costs. We derive three discrete choice models: the stochastic route choice model, the gravity type model for choice of origin, destination and route, a model for simultaneous choice of origin, destination, mode and route. The corresponding continuous models are derived.

8.1 Introduction

In Part I we derived and studied transportation models with constant link costs under the assumption of cost-minimizing behavior. We shall now derive the corresponding models for flow dependent separable link costs, separable here meaning that the travel cost on a link depends only on the total flow on that link. We shall assume that link costs are nondecreasing functions of link flow. We derive models for integer flows and give the corresponding continuous approximations.

8.2 Smith's Cumulative Cost Function and Equilibrium

The essential step in studying equilibrium with flow dependent link costs is the replacement of constant link costs with the cumulative cost function.

Let there be T trip makers, each choosing a specific route $r_t, t = 1, \ldots, T$. A *trip pattern* $\mathbf{u}$ contains T routes,

$$\mathbf{u} = (r_1, \ldots, r_T).$$

Most quantities of interest will be functions of the trip pattern $\mathbf{u}$. Note that there will usually be many trip patterns differing only by the order of the trips. This is a fundamental property and we shall start by investigating some of the consequences.

Let $c(r_t|r_1, \ldots, r_{t-1})$ be the cost to trip maker t if he chooses route r_t when the first trip makers have chosen routes $(r_1, \ldots, r_{t-1})$. Let $c(\mathbf{u})$ denote the *cumulative*

S. Erlander, *Cost-Minimizing Choice Behavior in Transportation Planning*, Advances in Spatial Science, DOI 10.1007/978-3-642-11911-8_8,

user cost function defined by

$$c(\mathbf{u}) = \sum_{t=1}^{T} c(r_t | r_1, \ldots, r_{t-1}) = \sum_{t=1}^{T} \sum_{l \in r_t} c_l(v_l^t), \tag{8.1}$$

where v_l^t denotes the volume on link l when the first t trips $r_1, \ldots, r_t$ have been allocated and where $c_l(v_l^t)$ is the travel cost on link l incurred by a trip maker who chooses link l when there are already $(v_l^t - 1)$ trip makers using link l. Let $v_l(\mathbf{u})$ be the volume on link l when all trip makers have been allocated. The cumulative cost function will permit a very nice formulation of user equilibrium.

We do not permit routes having a loop. This means that each route may use a specific link once only. This observation leads to the following proposition.

Proposition 22.

$$c(\mathbf{u}) = \sum_{l} \sum_{k=1}^{v_l(\mathbf{u})} c_l(k). \tag{8.2}$$

Proof. According to definition

$$c(\mathbf{u}) = \sum_{t=1}^{T} \sum_{l \in r_t} c_l(v_l^t).$$

It is evident that we can sum (as here) over trips and for each trip over the links used, or over links and for each link over the load up to $v_l(\mathbf{u})$. In both cases the terms $c_l(k), k = 1, \ldots, v_l(\mathbf{u})$, will appear exactly once. □

Formula (8.2) shows that the order of the trips is irrelevant. Since the order of the trips is irrelevant, we have in particular

$$c(\mathbf{u}) = c(r_1, \ldots, r_t, \ldots, r_T) = c(r_1, \ldots, r_{t-1}, r_{t+1}, \ldots, r_T, r_t), \; t = 1, \ldots, T. \tag{8.3}$$

This means that every trip maker t may be considered as the last one.

Definition 19 (User Equilibrium). $\mathbf{u}^* = (r_1^*, \ldots, r_T^*)$ is a *user equilibrium* if and only if for every r_t with the same origin and destination as r_t^* we have

$$c(r_t^* | r_1^*, \ldots, r_{t-1}^*, r_{t+1}^*, \ldots, r_T^*) \leq c(r_t | r_1^*, \ldots, r_{t-1}^*, r_{t+1}^*, \ldots, r_T^*), \; t = 1, \ldots, T. \tag{8.4}$$

The definition implies that if $\mathbf{u}^*$ is a user equilibrium then it is not possible for any one of the trip makers, given his origin and destination, to take an alternate route that is less costly than the route r^* already chosen. Thus $\mathbf{u}^*$ is a Nash equilibrium. In fact, if $\mathbf{u}^*$ is a minimum value of the cumulative cost function, then it is a user equilibrium:

Proposition 23.

$$c(\mathbf{u}^*) = \min c(\mathbf{u}) \implies \mathbf{u}^* \textit{ is a user equilibrium.}$$

Proof. Let $\mathbf{u}^*$ be a solution to the minimization problem $\min c(\mathbf{u})$. We wish to show that (8.4) is satisfied for every $t = 1, \ldots, T$. Assume that there is some k such that (8.4) is not satisfied for this k. Since the order of the routes is irrelevant, let k be the last route. Then replacing r_k^* with r_k in $\mathbf{u}^*$ the so obtained trip pattern has smaller value for the cumulative cost function. But $\mathbf{u}^*$ was assumed to be optimal. To remove the contradiction we must conclude that $\mathbf{u}^*$ is a user equilibrium. □

There is a finite number of trip patterns. Hence, the trip patterns can be enumerated, $\mathbf{u}_1, \ldots, \mathbf{u}_K$, where K is a usually very large number. The choice set for the trip makers is $[\mathbf{u}_1, \ldots, \mathbf{u}_K]$.

The minimum in Proposition 23 is obtained for some $k \in (1, \ldots, K)$,

$$c(u^*) = \min c(\mathbf{u}) = c(\mathbf{u}_k) \text{ for some } k \in (1, \ldots, K).$$

8.3 Route Choice

In Sect. 5.2 we derived the stochastic route choice model with constant link costs (5.2). We shall now derive the corresponding model for flow dependent separable link costs.We shall assume that the decisions by the trip makers can be described by a probability distribution $p = (p_1, \ldots, p_K)$, $p_k = \Pr(\mathbf{u} = \mathbf{u}_k)$.

If the behavior of the trip makers is cost minimizing, then the last trip maker will choose a route r_T that minimizes the last term of the cumulative cost function $c(r_T | r_1, \ldots, r_{T-1})$. However, according to (8.3) every trip maker t may be considered as the last one. Thus every trip maker will choose a route that minimizes his cost given the choices of all other trip makers. Hence, if cost-minimizing behavior prevails, we would expect the cumulative cost function $c(\mathbf{u})$ to take lower values more frequently than higher values.

Let there be two independent sequences of trip patterns, $\mathbf{u}^1, \ldots, \mathbf{u}^N$ and $\mathbf{w}^1, \ldots, \mathbf{w}^N$. Let z_k^1 = the number of times the particular trip pattern $\mathbf{u}_k$ appears in the sequence $\mathbf{u}^1, \ldots, \mathbf{u}^N$. Let z_k^2 = be the corresponding number in the sequence $\mathbf{w}^1, \ldots, \mathbf{w}^N$.

Cost-minimizing behavior is defined as in Chap. 4, with the value of the cumulative cost function $c(\mathbf{u}_k)$ replacing the cost c_k.

Definition 20 (Cost-Minimizing Behavior). The probability distribution p is a cost-minimizing probability distribution with respect to the cumulative cost function

$c(\mathbf{u})$ iff for every pair of independent samples $\mathbf{u}^1, \ldots, \mathbf{u}^N$ and $\mathbf{w}^1, \ldots, \mathbf{w}^N$ and for any N we have

$$c(\mathbf{u}^1) + \cdots + c(\mathbf{u}^N) \leq c(\mathbf{w}^1) + \cdots + c(\mathbf{w}^N) \Longrightarrow \Pr(\mathbf{u}^1, \ldots, \mathbf{u}^N) \geq \Pr(\mathbf{w}^1, \ldots, \mathbf{w}^N),$$

or equivalently

$$\sum_{k=1}^{K} z_k^1 c(\mathbf{u}_k) \leq \sum_{k=1}^{K} z_k^2 c(\mathbf{w}_k) \Longrightarrow \prod_{k=1}^{K} p_k^{z_k^1} \geq \prod_{k=1}^{K} p_k^{z_k^2}.$$

The definition of cost-minimizing behavior means that when taking a series of independent samples $\mathbf{u}^1, \ldots, \mathbf{u}^N$ we expect to observe lower values of the sum of cumulative cost functions $(c(\mathbf{u}^1) + \cdots + c(\mathbf{u}^N))$ more frequently than higher values. In the same way as in Chap. 4 we obtain the following proposition.

Proposition 24 (Cost-Minimizing Behavior). *The probability distribution* $p = (p_1, \ldots, p_K)$ *represents cost-minimizing behavior in the sense of Definition 20 if and only if it is a log linear probability distribution* $p_k = \exp(\mu - \gamma c(\mathbf{u}_k))$ *which can equivalently be written*

$$p_k = \frac{\exp(-\gamma c(\mathbf{u}_k))}{\sum_{k=1}^{K} \exp(-\gamma c(\mathbf{u}_k))}, \tag{8.5}$$

where $\gamma \geq 0$.

Proof. The proposition follows from Proposition 1 if we identify c_k with the cumulative cost function $c(\mathbf{u}_k)$, for $k = 1, \ldots, K$. □

If the links costs are independent of flow, $c_l(k) = c_l$, the cumulative cost function becomes

$$c(\mathbf{u}) = \sum_{l} \sum_{k=1}^{v_l(\mathbf{u})} c_l(k) = \sum_{l} v_l(\mathbf{u}) c_l,$$

which can be identified as the total cost function (4.1) used in Sect. 4.2. For constant link costs we obtain the stochastic route choice model for constant link costs (5.2) given in Sect. 5.2.

We are now ready to give the remarkable result that the most probable trip patterns are user equilibria, so called stochastic user equilibria, SUE, or dispersed equilibria.

Proposition 25 (Most Probable Trip Patterns are User Equilibria). *Let cost-minimizing behavior according to Definition 20 prevail. Then the most probable trip patterns are user equilibria.*

Proof. The most probable trip patterns are obtained by simply minimizing the cumulative cost function in (8.5),

$$\max_k p_k = \max_k \frac{\exp(-\gamma c(\mathbf{u}_k))}{\sum_{k=1}^{K} \exp(-\gamma c(\mathbf{u}_k))} = \frac{\exp(-\gamma c(\mathbf{u}^*))}{\sum_{k=1}^{K} \exp(-\gamma c(\mathbf{u}_k))},$$

where $\mathbf{u}^* = \min_k c(\mathbf{u}_k)$. Since $\mathbf{u}^*$ is a minimum it follows from Proposition 23 that it is a user equilibrium. Hence the most probable trip patterns are user equilibria. □

8.3.1 Continuous Approximation

One way of solving the maximization problem approximately is by relaxing the restriction to integer values.

Let v_r be the flow on route r from one place to another place and let $\mathbf{v} = [v_r]$. We have

$$v_l = \sum_r v_r \delta_{lr},$$

where

$$\delta_{lr} = 1 \text{ if route } r \text{ uses link } l, \; 0 \text{ otherwise.}$$

We shall replace the cumulative cost function

$$c(\mathbf{u}) = \sum_l \sum_{k=1}^{v_l(\mathbf{u})} c_l(k)$$

with the Beckmann integral

$$\mathcal{C}(\mathbf{v}) = \sum_l \int_0^{v_l} c_l(v)\mathrm{d}v. \tag{8.6}$$

Proposition 26.

$$\mathcal{C}(\mathbf{v}^*) = \min \mathcal{C}(\mathbf{v}) \Rightarrow \mathbf{v}^* \textit{ is a user equilibrium.}$$

Proof. Consider the optimization problem

$$\min \; \mathcal{C}(\mathbf{v}) \text{ subject to } \sum_r v_r = T.$$

This problem has the Lagrangian

$$\mathcal{L} = \sum_l \int_0^{v_l} c_l(v)\mathrm{d}v + \lambda(T - \sum_r v_r).$$

Differentiation with respect to v_r gives

$$\frac{\delta \mathcal{L}}{\delta v_r} = \sum_l c_l(v_l)\delta_{lr} - \lambda,$$

for all r. Hence, the optimum conditions are equal to Wardrop's user equilibrium conditions,

$$v_r^* > 0 \implies \sum_l c_l(v_l)\delta_{lr} = \lambda,$$

$$v_r^* = 0 \implies \sum_l c_l(v_l)\delta_{lr} \geq \lambda,$$

and v_r^* is a user equilibrium. This means that at optimum all used routes have the same cost λ and satisfy the Wardrop user equilibrium conditions. □

8.4 Choice of Origin, Destination and Route

We are now going to extend the previous results to the simultaneous choice of origin, destination and route. We have a network with a set of origins, $i = 1, \ldots, I$, a set of destinations, $j = 1, \ldots, J$, and a set of directed links $\mathcal{L}$ connecting the origins with the destinations. Suppose there are T trip makers, each choosing a specific route $r_t, t = 1, \ldots, T$. A *trip pattern* $\mathbf{u}$ contains T routes, $\mathbf{u} = (r_1, \ldots, r_T)$, where each route $r_t = r$ travels over some links connecting origins and destinations.

There is a finite number of trip patterns. Hence, the trip patterns can be enumerated, $\mathbf{u}_1, \ldots, \mathbf{u}_K$, where K is a usually very large number. The choice set is $[\mathbf{u}_1, \ldots, \mathbf{u}_K]$.

In defining cost-minimizing behavior we restrict comparisons to trip patterns which are activity equivalent.

Let $v_{ijr}(\mathbf{u})$ = the number of trips in the trip pattern $\mathbf{u}$ from origin i to destination j using route r. We define

$$T_{ij}(\mathbf{u}) = \sum_r v_{ijr}(\mathbf{u}), \quad A_i(\mathbf{u}) = \sum_j T_{ij}(\mathbf{u}), \quad B_j(\mathbf{u}) = \sum_i T_{ij}(\mathbf{u}).$$

$A_i(\mathbf{u})$ and $B_j(\mathbf{u})$ can be identified as the marginal totals of the trip matrix $[T_{ij}(\mathbf{u})]$. When defining cost-minimizing behavior we restrict attention to trip patterns with the same marginal totals.

Let $\mathbf{a}(\mathbf{u})$ denote the vector of all marginal totals,

$$\mathbf{a}(\mathbf{u}) = (A_1(\mathbf{u}), \ldots, A_I(\mathbf{u}),\ B_1(\mathbf{u}), \ldots, B_J(\mathbf{u})).$$

Keeping this activity vector constant, $\mathbf{a}(\mathbf{u}) = \mathbf{a}$, gives the following result corresponding to Proposition 23.

Proposition 27.

$$c(\mathbf{u}^*) = \min c(\mathbf{u}) \textit{ subject to } \mathbf{a}(\mathbf{u}) = \mathbf{a} \quad \Rightarrow \quad \mathbf{u}^* \textit{ is a user equilibrium.}$$

Proof. The proof parallels the proof of Proposition 23. □

In Proposition 25 we showed that for the route choice problem the most probable trip patterns are user equilibria. We shall now extend this property to the choice of origin, destination and route.

8.4.1 Most Probable Trip Pattern

Let there be two independent sequences of trip patterns, $\mathbf{u}^1, \ldots, \mathbf{u}^N$ and $\mathbf{w}^1, \ldots, \mathbf{w}^N$. Let z_k^1 = the number of times the particular trip pattern $\mathbf{u}_k$ appears in the sequence $\mathbf{u}^1, \ldots, \mathbf{u}^N$. Let z_k^2 = be the corresponding number in the sequence $\mathbf{w}^1, \ldots, \mathbf{w}^N$.

Activity equivalence is now defined in the following way.

Definition 21 (Activity Equivalence). The two trip pattern samples $\mathbf{u}^1, \ldots, \mathbf{u}^N$ and $\mathbf{w}^1, \ldots, \mathbf{w}^N$ are activity equivalent if

$$\sum_{k=1}^{K} z_k^1 A_i(\mathbf{u}_k) = \sum_{k=1}^{K} z_k^2 A_i(\mathbf{w}_k), i = 1, \ldots, I,$$

and

$$\sum_{k=1}^{K} z_k^1 B_j(\mathbf{u}_k) = \sum_{k=1}^{K} z_k^2 B_j(\mathbf{w}_k), j = 1, \ldots, J. \tag{8.7}$$

Activity equivalence means that we compare only samples $\mathbf{u}^1, \ldots, \mathbf{u}^N$ and $\mathbf{w}^1, \ldots, \mathbf{w}^N$ having equal total number of trips going from zones $i = 1, \ldots, I$ to zones $j = 1, \ldots, J$.

The cumulative cost function is defined as before, (8.1).

Definition 22 (Cost-Minimizing Behavior). The probability distribution p is a cost-minimizing probability distribution with respect to the cumulative cost function $c(\mathbf{u})$ iff for every pair of activity equivalent independent samples $\mathbf{u}^1, \ldots, \mathbf{u}^N$ and $\mathbf{w}^1, \ldots, \mathbf{w}^N$ and for any N we have

$$[\sum_{k=1}^{K} z_k^1 c(\mathbf{u}_k) \leq \sum_{k=1}^{K} z_k^2 c(\mathbf{w}_k)] \Longrightarrow [\prod_{k=1}^{K} p_k^{z_k^1} \geq \prod_{k=1}^{K} p_k^{z_k^2}].$$

The content of this definition of cost-minimizing behavior is that when taking a series of independent samples $\mathbf{u}^1, \ldots, \mathbf{u}^N$ we expect to observe lower values of the sum of cumulative cost functions $(c(\mathbf{u}^1) + \cdots + c(\mathbf{u}^N))$ more frequently than higher values.

Again, in the same way as in Chap. 4 we obtain the following proposition.

Proposition 28 (The Representation of Cost-Minimizing Behavior). *The probability distribution $p = (p_1, \ldots, p_K)$ represents cost-minimizing behavior in the sense of Definition 22 if and only if for some real numbers α_i, β_j and γ*

$$p_k = \exp[\sum_i \alpha_i A_i(\mathbf{u}_k) + \sum_j \beta_j B_j(\mathbf{u}_k) - \gamma c(\mathbf{u}_k)], \tag{8.8}$$

where $\gamma \geq 0$.

Proof. The proposition follows from Proposition 3, where the matrix A with elements $[A_i(\mathbf{u}_k)]$ and $[B_j(\mathbf{u}_k)]$ is given by $\mathrm{A} = [\mathbf{a}(\mathbf{u}_1)^T \cdots \mathbf{a}(\mathbf{u}_K)^T]$, and where the cost coefficients c_k are replaced by $c(\mathbf{u}_k)$. □

In Proposition 25 we showed that for the route choice problem the most probable trip patterns are user equilibria. We shall now extend this property to the choice of origin, destination and route.

Proposition 29 (Most Probable Trip Patterns are User Equilibria). *Let cost-minimizing behavior according to Definition 22 prevail. Then the most probable trip patterns are user equilibria.*

Proof. The most probable trip patterns are obtained by maximizing the probability (8.8). For any $\mathbf{a}$, consider trip patterns $\mathbf{u}_k$ satisfying $\mathbf{a}(\mathbf{u}_k) = \mathbf{a}$. For these trip patterns the marginal totals $A_i(\mathbf{u}_k)$ and $B_j(\mathbf{u}_k)$ are constant. Hence maximum of the probability is obtained simply by minimizing the cumulative cost function subject to $\mathbf{a}(\mathbf{u}_k) = \mathbf{a}$,

$$c(\mathbf{u}^*) = \min_k c(\mathbf{u}_k) \quad \text{subject to} \quad \mathbf{a}(\mathbf{u}_k) = \mathbf{a}. \tag{8.9}$$

The optimal solution $\mathbf{u}^*$ to (8.9) is a most probable trip pattern given that $\mathbf{a}(\mathbf{u}_k) = \mathbf{a}$. It then follows from Proposition 27 that $\mathbf{u}^*$ is a user equilibrium. Hence, for any $\mathbf{a}$, $\mathbf{u}^*$ is a most probable trip pattern and at the same time a user equilibrium. Thus the most probable trip patterns are user equilibria. □

It is a remarkable fact that cost-minimizing behavior implies that the most probable trip patterns are user equilibria.

Note that in practical applications the parameters (multipliers) α_i, β_j and γ have to be estimated. Usually the marginal totals will be given by the problem, $\mathbf{a}(\mathbf{u}) = \mathbf{a}$, and it remains to find the optimal solution to the minimization problem in (8.9).

8.4.2 Most Probable Flow

We shall now study the flow patterns $\mathbf{v} = [v_{ijr}(\mathbf{u})]$. Let $\mathbf{v} = [v_{ijr}]$ be a specific flow pattern. Clearly, v_{ijr} is a function of $\mathbf{u}$, $v_{ijr} = v_{ijr}(\mathbf{u})$, and many trip patterns $\mathbf{u}$ correspond to the same flow pattern v_{ijr}. Let U_v be the set of all trip patterns $\mathbf{u}$ that correspond to this specific flow pattern, $U_v = [\mathbf{u} \mid [\mathbf{v}_{ijr}(\mathbf{u})] = [v_{ijr}]]$. The number of elements in U_v is $T!/\prod_{i,j,r}(v_{ijr}!)$. Let

$$\delta_{ij,lr} = \begin{cases} 1 \text{ if route } r \text{ from } i \text{ to } j \text{ uses link } l, \\ 0 \text{ otherwise.} \end{cases}$$

Proposition 30 (Flow Pattern Probability). *Let cost-minimizing behavior prevail. Then the probability* $\Pr(\mathbf{v})$ *of the specific flow pattern* $\mathbf{v} = [v_{ijr}]$ *is given, for* $\gamma \geq 0$, *and for any* $\mathbf{u} \in U_v$, *by*

$$\Pr(\mathbf{v}) = \frac{T!}{\prod_{i,j,r}(v_{ijr}!)} \exp[\sum_i \alpha_i A_i(\mathbf{u}) + \sum_j \beta_j B_j(\mathbf{u}) - \gamma c(\mathbf{u})]. \tag{8.10}$$

Proof. Consider a trip pattern $\mathbf{u} \in U_v$. The probability $\Pr(\mathbf{u})$ of this trip pattern is given by Proposition 28. Furthermore

$$T_{ij}(\mathbf{u}) = \sum_{r=1}^{R} v_{ijr}, \; A_i(\mathbf{u}) = \sum_{j=1}^{J} T_{ij}(\mathbf{u}), \; B_j(\mathbf{u}) = \sum_{i=1}^{I} T_{ij}(\mathbf{u}).$$

Hence $T_{ij}(\mathbf{u})$, $A_i(\mathbf{u})$ and $B_j(\mathbf{u})$ are constant for all $\mathbf{u} \in U_v$. Similarly, $v_l(\mathbf{u})$ is the load on link l when all trips in $\mathbf{u}$ have been allocated,

$$v_l(\mathbf{u}) = \sum_{i,j,r} v_{ijr}(\mathbf{u})\delta_{ij,lr}, \tag{8.11}$$

where $v_{ijr}(\mathbf{u}) = v_{ijr}$ for $\mathbf{u} \in U_v$ and hence constant. Using (8.2) it follows that $c(\mathbf{u}) = \sum_l \sum_{k=1}^{v_l(\mathbf{u})} c_l(k)$ is constant for all $\mathbf{u} \in U_v$. Hence $\Pr(\mathbf{u})$ is constant for $\mathbf{u} \in U_v$. To obtain $\Pr(\mathbf{v})$ we simply have to multiply with the number of elements $\mathbf{u} \in U_v$. This gives the formula in the proposition. □

The most probable flow patterns can be obtained by maximizing the probability $\Pr(\mathbf{v})$ in Proposition 30. The trip pattern $\mathbf{u}$ does not appear explicitly in (8.10). The marginal totals $A_i(\mathbf{u})$ and $B_j(\mathbf{u})$ depend only on $\mathbf{v}$. Also the cumulative cost function depends only on $\mathbf{v}$ because

$$v_l(\mathbf{u}) = \sum_{i,j,r} v_{ijr}(\mathbf{u})\delta_{ij,lr} = \sum_{i,j,r} v_{ijr}\delta_{ij,lr} = v_l.$$

Using (8.2) we obtain

$$c(\mathbf{u}) = \sum_l \sum_{k=1}^{v_l(\mathbf{u})} c_l(k) = \sum_l \sum_{k=1}^{v_l} c_l(k).$$

Hence the formula contains only the flow pattern $\mathbf{v} = [v_{ijr}]$, and we obtain:

Proposition 31 (Most Probable Flow Pattern). *Let cost-minimizing behavior prevail. Then the most probable flow patterns are given as the optimal solutions to the problem*

$$\max \quad \frac{T!}{\prod_{i,j,r}(v_{ijr}!)} \exp[\sum_{i,j,r} \alpha_i v_{ijr} + \sum_{i,j,r} \beta_j v_{ijr} - \gamma \sum_l \sum_{k=1}^{v_l} c_l(k)], \qquad (8.12)$$

subject to

$$\sum_{j,r} v_{ijr} = A_i, \ i = 1, \ldots, I, \ \sum_{i,r} v_{ijr} = B_j, \ j = 1, \ldots, J. \qquad (8.13)$$

Proof. We need only to express the quantities in (8.10) as functions of v_{ijr} in order to obtain (8.12). □

One way of solving the maximization problem approximately is by relaxing the restriction to integer values. This formulation will be given next.

8.4.3 Continuous Approximation

In previous sections (except for Sect. 8.3.1) we have treated each trip maker as one unit. This leads to integer values on flows. This has given several interesting results, mostly formulated in probabilistic terms. However, these results also have corresponding continuous deterministic versions. We treated the route choice problem in Sect. 8.3.1 We shall now give the continuous deterministic approximation of the combined origin, destination and route choice problem. In order to obtain the continuous counterparts we shall let the variables be real numbers and replace the relevant summations with Riemann integrals.

We are going to derive the deterministic continuous combined model for choices of origins, destinations and routes. The model will be derived by relaxing the integer constraint on the number of trips and replacing the cumulative cost function by the Beckmann integral. This will be an approximative model since the number of trips is treated as a real variable and not restricted to integer variables. To make this section more self contained we repeat some of the notation.

Let

$$\delta_{ij,lr} = \begin{cases} 1 \text{ if route } r \text{ from } i \text{ to } j \text{ uses link } l, \\ 0 \text{ otherwise.} \end{cases}$$

Furthermore, let v_{ijrl} be the flow from origin i to destination j using route r on link l. Then $v_{ijr} = \sum_l \delta_{ij,lr}$ denotes the flow from origin i to destination j using route r. Let $\mathbf{v} = [v_{ijr}]$. Let $v_l = \sum_{i,j,r} v_{ijr}\delta_{ij,lr}$.

Proposition 32 (Continuous Approximation of most Probable Flow Pattern). *Let the trip pattern distribution represent cost-minimizing behavior. Then the most probable flow patterns given the marginal totals are approximately given by the unique optimal solution* $\mathbf{v}^* = [v^*_{ijr}]$ *to the problem, where* $\gamma \geq 0$,

$$\min_{\mathbf{v}} \; [\sum_{i,j,r} v_{ijr}(\log v_{ijr} - 1) + \gamma \sum_l \int_0^{v_l} c_l(v)\mathrm{d}v] \;\; s.\,t.\; v_{ijr} \geq 0 \;\; for\; all\;\; i, j, r \tag{8.14}$$

$$and \; \sum_{j,r} v_{ijr} = A_i, \; i = 1, \ldots, I, \; \sum_{i,r} v_{ijr} = B_j. \; j = 1, \ldots, J. \tag{8.15}$$

Furthermore the optimal solution satisfies for some real numbers α_i *and* β_j

$$v^*_{ijr} = \exp(\alpha_i + \beta_j - \gamma c^*_{ijr}), \; where \; c^*_{ijr} = \sum_l c_l(v^*_l)\delta_{ij,lr}. \tag{8.16}$$

The optimal solution $\mathbf{v}^* = [v^*_{ijr}]$ *is a dispersed solution carrying flow on all routes.*

Proof. The most probable flow patterns were given by Proposition 31 as the optimal solutions to the problem,

$$\max \Pr(\mathbf{v}) = \frac{T!}{\prod_{i,j,r}(v_{ijr}!)} \exp[\sum_{i,j,r} \alpha_i v_{ijr} + \sum_{i,j,r} \beta_j v_{ijr} - \gamma \sum_l \sum_{k=1}^{v_l} c_l(k)].$$

Taking logarithms, neglecting constant terms, approximating the factorials by Stirling's formula and replacing the cumulative sum function by its continuous approximation, the Beckmann integral

$$\sum_l \int_0^{v_l} c_l(v)\mathrm{d}v,$$

we obtain the maximization problem

$$\max[-\sum_{i,j,r} v_{ijr}(\log v_{ijr}) - \gamma \sum_l \int_0^{v_l} c_l(v)\mathrm{d}v]. \tag{8.17}$$

Using the fact that $\max f = -min(-f)$, and adding the constant expression

$$\sum_{i,j,r} v_{ijr} = T,$$

(8.17) is seen to be equivalent to (8.14). Uniqueness follows from the strict concavity of the first sum in the objective function.

The Lagrangian to the problem of minimizing the objective function (8.14) subject to the marginal constraints (8.15) becomes

$$\begin{aligned}\mathcal{L}(\mathbf{v}) = &\sum_{i,j,r} v_{ijr}(\log v_{ijr} - 1) + \gamma \sum_l \int_0^{v_l} c_l(v)\mathrm{d}v \\ &+ \sum_i \alpha_i (A_i - \sum_{j,r} v_{ijr}) + \sum_j \beta_j (B_j - \sum_{i,r} v_{ijr}).\end{aligned}$$

It follows that

$$\frac{\delta \mathcal{L}}{\delta v_{ijr}} = \log v_{ijr} + \gamma \sum_l c_l(v_l)\delta_{ij,lr} - (\alpha_i + \beta_j).$$

The optimality conditions are

$$v^*_{ijr} > 0 \implies \log v_{ijr} + \gamma \sum_l c_l(v_l)\delta_{ij,lr} - (\alpha_i + \beta_j) = 0.$$

and (8.16) follows. □

Note that the optimization problem (8.14) and (8.15) is here obtained from cost-minimizing behavior as an approximation to the exact problem of finding the most probable flow patterns. The classic approach is to set up the optimization problem because the particular form of the solutions is wanted.

In a most probable flow pattern v^* – more properly – in an approximation to a most probable flow pattern all flow variables v^*_{ijr} have some flow, even if the flow in most relations is very close to zero.

The optimal solution to problem (8.14) is a gravity type model for flows. The parameters α_i and β_j are the Lagrange multipliers corresponding to the constraints (8.15).

8.5 Choice of Origin, Destination, Mode and Route

We shall now extend the previous results to the simultaneous choice of origin, destination, mode and route. This will be obtained simply by extending the definition of activity equivalence. We wish, in defining cost-minimizing behavior, to compare only trip patterns where, for each mode $m = 1, \ldots, M$, the total number of trip makers choosing the mode m is the same.

The proofs of the propositions in this section exactly parallels the proofs of the corresponding propositions in Sect. 8.4.

We have a network with a set of origins, $i = 1, \ldots, I$, a set of destinations, $j = 1, \ldots, J$, and a set of directed links $\mathcal{L}$ connecting the origins with the destinations. Suppose there are T trip makers, each choosing a specific route $r_t, t = 1, \ldots, T$. A *trip pattern* $\mathbf{u}$ contains T routes, $\mathbf{u} = (r_1, \ldots, r_T)$, where each route $r_t = r$ travels over some links connecting origins and destinations. The links represent different modes of travel.

There is a finite number of trip patterns. Hence, the trip patterns can be enumerated, $\mathbf{u}_1, \ldots, \mathbf{u}_K$, where K is a usually very large number. The choice set is $[\mathbf{u}_1, \ldots, \mathbf{u}_K]$.

In defining cost-minimizing behavior we restrict comparisons to trip patterns which are activity equivalent.

Let $v_{ijmr}(\mathbf{u})$ = the number of trips in the trip pattern $\mathbf{u}$ from origin i to destination j on mode m using route r.

We define

$$T_{ij}(\mathbf{u}) = \sum_{m,r} v_{ijmr}(\mathbf{u}), \;\; A_i(\mathbf{u}) = \sum_j T_{ij}(\mathbf{u}),$$
$$B_j(\mathbf{u}) = \sum_i T_{ij}(\mathbf{u}), \;\; v_m(\mathbf{u}) = \sum_{i,j,r} v_{ijmr}(\mathbf{u}).$$

$A_i(\mathbf{u})$ and $B_j(\mathbf{u})$ can be identified as the marginal totals of the trip matrix $[T_{ij}(\mathbf{u})]$. $v_m(\mathbf{u})$ is the number of trips using mode m. When defining cost-minimizing behavior we restrict attention to trip patterns with the same marginal totals.

Let $\mathbf{a}(\mathbf{u})$ denote the vector of all marginal totals and total flows on modes,

$$\mathbf{a}(\mathbf{u}) = (A_1(\mathbf{u}), \ldots, A_I(\mathbf{u}), \; B_1(\mathbf{u}), \ldots, B_J(\mathbf{u}), \; v_m(\mathbf{u}), \; m = 1, \ldots, M).$$

Keeping this activity vector constant, $\mathbf{a}(\mathbf{u}) = \mathbf{a}$, gives the following result corresponding to Proposition 23.

Proposition 33.

$$c(\mathbf{u}^*) = \min c(\mathbf{u}) \textit{ subject to } \mathbf{a}(\mathbf{u}) = \mathbf{a} \;\; \Rightarrow \;\; \mathbf{u}^* \textit{ is a user equilibrium.}$$

Proof. The proof parallels the proof of Proposition 23. □

Let there be two independent sequences of trip patterns, $\mathbf{u}^1, \ldots, \mathbf{u}^N$ and $\mathbf{w}^1, \ldots, \mathbf{w}^N$. Let z_k^1 = the number of times the particular trip pattern $\mathbf{u}_k$ appears in the sequence $\mathbf{u}^1, \ldots, \mathbf{u}^N$. Let z_k^2 = be the corresponding number in the sequence $\mathbf{w}^1, \ldots, \mathbf{w}^N$.

Activity equivalence is now defined in the following way.

Definition 23 (Activity Equivalence). The two trip pattern samples $\mathbf{u}^1, \ldots, \mathbf{u}^N$ and $\mathbf{w}^1, \ldots, \mathbf{w}^N$ are activity equivalent if

$$\sum_k z_k^1 A_i(\mathbf{u}_k) = \sum_k z_k^2 A_i(\mathbf{w}_k),\ i = 1, \ldots, I,$$
$$\sum_k z_k^1 B_j(\mathbf{u}_k) = \sum_k z_k^2 B_j(\mathbf{w}_k),\ j = 1, \ldots, J$$

and

$$\sum_k z_k^1 v_m(\mathbf{u}_k) = \sum_k z_k^2 v_m(\mathbf{w}_k),\ m = 1, \ldots, M.$$

The cumulative cost function is defined as before, (8.1).

Definition 24 (Cost-Minimizing Behavior). The probability distribution p is a cost-minimizing probability distribution with respect to the cumulative cost function $c(\mathbf{u})$ iff for every pair of activity equivalent independent samples $\mathbf{u}^1, \ldots, \mathbf{u}^N$ and $\mathbf{w}^1, \ldots, \mathbf{w}^N$ and for any N we have

$$[\sum_k z_k^1 c(\mathbf{u}_k) \le \sum_k z_k^2 c(\mathbf{w}_k)] \Longrightarrow [\prod_k p_k^{z_k^1} \ge \prod_k p_k^{z_k^2}].$$

Proposition 34 (Cost-Minimizing Behavior). *The probability distribution $p = (p_1, \ldots, p_K)$ represents cost-minimizing behavior in the sense of Definition 24 if and only if for some real numbers α_i, β_j, μ_m and γ*

$$p_k = \exp[\sum_i \alpha_i A_i(\mathbf{u}_k) + \sum_j \beta_j B_j(\mathbf{u}_k) + \sum_m \mu_m v_m(\mathbf{u}_k) - \gamma c(\mathbf{u}_k)],$$

where $\gamma \ge 0$.

Proof. The proof parallels the proof of Proposition 28.

Proposition 35 (Most Probable Trip Patterns are User Equilibria). *Let cost-minimizing behavior according to Definition 24 prevail. Then the most probable trip patterns are user equilibria.*

Proof. The proof parallels the proof of Proposition 29.

For a specific trip pattern $[v_{ijmr}]$, let

$$U_v = [\mathbf{u} \mid [\mathbf{v}_{ijmr}(\mathbf{u})] = [v_{ijmr}]].$$

Proposition 36 (Flow Pattern Probability). *Let cost-minimizing behavior prevail. Then the probability* $\Pr(\mathbf{v})$ *of the specific flow pattern* $\mathbf{v} = [v_{ijmr}]$ *is given, for* $\gamma \geq 0$, *and for any* $\mathbf{u} \in U_v$, *by*

$$\Pr(\mathbf{v}) = \frac{T!}{\prod_{i,j,m,r}(v_{ijmr}(\mathbf{u})!)} \exp[\sum_i \alpha_i A_i(\mathbf{u}) + \sum_j \beta_j B_j(\mathbf{u}) + \sum_m \mu_m v_m(\mathbf{u}) - \gamma c(\mathbf{u})].$$

Proof. The proof parallels the proof of Proposition 30.

Proposition 37 (Most Probable Flow Pattern). *Let cost-minimizing behavior prevail. Then the most probable flow patterns are given as the optimal solutions to the problem*

$$\max \Pr(\mathbf{v})$$

subject to

$$\sum_{j,m,r} v_{ijmr} = A_i, \; i = 1, \ldots, I, \; \sum_{i,m,r} v_{ijmr} = B_j, \; j = 1, \ldots, J,$$

$$\sum_{i,j,r} v_{ijmr} = v_m, \; m = 1, \ldots, M.$$

Proof. The proof parallels the proof of Proposition 31.

8.5.1 Continuous Approximation

Let

$$\delta_{ijm,lr} = \begin{cases} 1 \text{ if route } r \text{ from } i \text{ to } j \text{ on mode } m \text{ uses link } l, \\ 0 \text{ otherwise,} \end{cases}$$

and let $v_{ijmrl}(\mathbf{u})$ be the flow from i to j on mode m on route r using link l. Let, furthermore, the flow on link l be defined by

$$v_l(\mathbf{u}) = \sum_{i,j,m,r} v_{ijmlr}(\mathbf{u})\delta_{ijm,lr},$$

and

$$c_{ijmr}(\mathbf{u}) = \sum_l c_l(v_l(\mathbf{u}))\delta_{ijm,lr}.$$

As before we shall relax the integer constraint on the flow variables and replace the relevant summations with Riemann integrals in order to obtain the continuous deterministic approximation of the combined origin, destination and route choice problem.

Proposition 38 (Continuous Approximation of most Probable Flow Pattern). *Let the trip pattern distribution represent cost-minimizing behavior. Then the most* *probable flow patterns given the marginal totals are approximately given by the unique optimal solution* $\mathbf{v}^* = [v^*_{ijmr}]$ *to the problem, where* $\gamma \geq 0$,

$$\min_{i,j,m,r} \left[\sum_{i,j,m,r} v_{ijmr}(\log v_{ijmr} - 1) + \gamma \sum_l \int_0^{v_l} c_l(v)dv \right],$$

subject to $v_{ijmr} \geq 0$, *for all* i, j, m, r *and*

$$\sum_{j,m,r} v_{ijmr} = A_i, \; i = 1, \ldots, I, \; \sum_{i,m,r} v_{ijmr} = B_j, \; j = 1, \ldots, J,$$
$$\sum_{i,j,r} v_{ijmr} = v_m, \; m = 1, \ldots, M.$$

Furthermore the optimal solution satisfies for some real numbers α_i, β_j *and* μ_m

$$v^*_{ijmr} = \exp(\alpha_i + \beta_j + \mu_m - \gamma c^*_{ijmr}),$$

where

$$c^*_{ijmr} = \sum_l c_l(v^*_l)\delta_{ijm,lr}.$$

The optimal solution $\mathbf{v}^* = [v^*_{ijmr}]$ *is a dispersed solution carrying flow on all modes and all routes.*

Proof. The proof parallels the proof of Proposition 32.

8.6 Comments and Extensions

The optimization problems (8.14) and (8.15) are here obtained from cost-minimizing behavior as an approximation to the exact problem of finding the most probable flow patterns. Models of this kind are often formulated because of a desire to obtain a solution similar to (8.16), see e.g. Evans (1976), Boyce (1980) and Fisk (1980). For an overview see Patriksson (1994).

The model in Proposition 32 is a continuous dispersed equilibrium in contrast to Evans' CDA-model (Evans 1976), since here all routes contain some flow, even if the flow is very small on most routes, whereas in the CDA-model shortest routes, only, have any flow [Wardrop equilibrium (Wardrop 1952)].

The distribution for the most probable trip pattern, (8.8) and the following development, can be extended to different problem situations by adding constraints to the definition of activity equivalence (8.7). In this way models such as e.g. the models treated in Boyce et al. (1983, 1988), Boyce (1984) and Abrahamsson and Lundqvist (1999) can be obtained.

An algorithm for solving the optimization problem (8.14) has been proposed by Lundgren and Patriksson (1998).

8.7 Notes

The material in this Chapter was obtained by T.E. Smith for the *cost-efficiency principle* introduced by him (Smith 1978a,b) and further investigated in Smith (1983, 1988). Our starting point is *cost-minimizing behavior*, which is a slight reformulation of the cost-efficiency principle. Proposition 23 is Theorem 5 in Smith (1983). Proposition 32 is Theorem 3 in Erlander (1990).

The notion of activity equivalence as used in this chapter is the same as "strong efficiency" in Smith (1983). The term "dispersed equilibrium" was used by Smith (1988).

Rosenthal (1973) used the discrete analog (8.2) to the Beckmann integral (8.6) to derive equilibrium in a game theoretic setting.

Chapter 9
Appendix

Abstract We give some theorems on the general form of the Representation Theorem for cost-minimizing probability distributions, on maximum entropy and on maximum likelihood.

9.1 Representation Theorem for Cost-Minimizing Probability Distributions

We shall here give the general form of the Representation Theorem for cost-minimizing probability distributions. Cost-minimizing probability distributions were studied under the name of efficient probability distributions by Erlander and Smith (1990). The notion of efficiency is due to Smith (1978a,b). We need some notation and definitions.

Let the *activity matrix* be $\mathrm{A} \in \mathrm{R}^{M \times K}$ and let there be a *cost matrix* $\mathrm{C} \in \mathrm{R}^{S \times K}$. Let p_k be the probability of decision k, $k = 1, \ldots K$,

$p = (p_1, \ldots, p_K)^T \in \mathrm{P}_{++}^K$.

Let z_k be the number of times decision k is taken in an independent sample of size N from the probability distribution $[p_k]$. Let z denote the vector $(z_1, \ldots, z_K)^T$, $z \in \mathrm{Z}_+^K$. Let z'_k be number of times decision k is chosen in another independent sample of the same size N, and let z' denote the corresponding vector. Define the unit vector in R^K, $\kappa = (1, \ldots, 1)^T$.

Definition 25 (Cost-Minimizing Behavior). For any given activity matrix $\mathrm{A} \in \mathrm{R}^{M \times K}$ and cost matrix $\mathrm{C} \in \mathrm{R}^{S \times K}$, a positive probability distribution p is said to represent cost-minimizing behavior if and only if for all independent samples z and z' of the same size N

$$\mathrm{A}z = \mathrm{A}z', \mathrm{C}z \leq \mathrm{C}z' \quad \text{implies} \quad \prod_{k=1}^{K} p_k^{z_k} \geq \prod_{k=1}^{K} p_k^{z'_k}.$$

S. Erlander, *Cost-Minimizing Choice Behavior in Transportation Planning*,
Advances in Spatial Science, DOI 10.1007/978-3-642-11911-8_9,

Definition 26 (C-Identifiability).

- For any matrices, $\mathrm{A} \in \mathrm{R}^{M\times K}$ and $\mathrm{C} \in \mathrm{R}^{S\times K}$ the pair (A, C) is said to be C-IDENTIFIABLE if and only if for each column of the transpose matrix, $\mathrm{C}^T = [c_1, \ldots, c_S]$,

$$c_i \notin span([\kappa, \mathrm{A}^T, c_1, \ldots, c_{i-1}, c_{i+1}, \ldots, c_S]).$$

- If in addition the matrix, $[\kappa, \mathrm{A}^T, \mathrm{C}^T] \in \mathrm{R}^{K\times(1+M+S)}$ is of full column rank, then (A, C) is said to be FULLY IDENTIFIABLE.

Now, letting $\mathrm{Q}^{M\times K}$ denote the subset of rational-valued matrices in $\mathrm{R}^{M\times K}$ and writing $exp(x) = [exp(x_1), \ldots, exp(x_K)]$, the Representation Theorem can be formulated (Erlander and Smith (1990):

Theorem 1 (Representation Theorem for Cost-Minimizing Probability Distributions).

- *For any matrices,* $\mathrm{A} \in \mathrm{Q}^{M\times K}$ *and* $\mathrm{C} \in \mathrm{R}^{S\times K}$, *and any positive probability distribution p, p represents cost-minimizing behavior if and only if there exists a nonnegative vector,* $\gamma \in \mathrm{R}^S$, *such that for some vector,* $\alpha \in \mathrm{R}^M$,

$$p = \exp(\mu\kappa + \mathrm{A}^T\alpha - \mathrm{C}^T\gamma), \tag{9.1}$$

where $\mu = -\log[\kappa^T \exp(\mathrm{A}^T\alpha - \mathrm{C}^T\gamma)]$.
- *If* (A, C) *is C-identifiable then* γ *is unique,*
- *If* (A, C) *is fully identifiable, then* α *is also unique.*

Proof. This is the Representation Theorem (Theorem 2.1) in Erlander and Smith (1990). □

The geometric content of Theorem 1 is that the vector $-\log p$ must be contained in the polar cone spanned by the normals to the hyper planes defined by $[\kappa z = \kappa z' = N, \mathrm{A}z = \mathrm{A}z', \mathrm{C}z \le \mathrm{C}z']$. Identifiability is essential for uniqueness.

Probability distributions of the log linear form (9.1) are uniquely defined by their expected values as can be seen from the next theorem.

Let V denote the set of all matrix-vector pairs $(\mathrm{A}, \mathrm{C}, a, c) \in \mathrm{R}^{M\times K} \times \mathrm{R}^{S\times K} \times \mathrm{R}^I \times \mathrm{R}^S$ satisfying the two conditions

$$\text{the matrix B} = \begin{Bmatrix} 1 \cdots 1 \\ \mathrm{A} \\ \mathrm{C} \end{Bmatrix} \text{ is of full rank, and}$$

$$\begin{Bmatrix} 1 \cdots 1 \\ A \\ C \end{Bmatrix} p = \begin{Bmatrix} 1 \\ a \\ c \end{Bmatrix},$$

for some positive probability distribution p.

Theorem 2 (Uniqueness).

- *On the set V there exist unique functions $p : V \to \mathrm{P}^K_{++}$, $\mu : V \to \mathrm{R}$, $\alpha : V \to \mathrm{R}^M$, and $\gamma : V \to \mathrm{R}^S$ such that for all $v = (\mathrm{A}, \mathrm{C}, a, c) \in V$,*

$$\begin{aligned} \mathrm{A}p(v) &= a, \\ \mathrm{C}p(v) &= c, \\ p(v) &= \exp(\mu(v)\kappa + \mathrm{A}^T\alpha(v) - \mathrm{C}^T\gamma(v)), \end{aligned}$$

  where $\mu(v) = -\log(\kappa^T \exp[\mathrm{A}^T\alpha(v) - \mathrm{C}^T\gamma(v)])$ holds for all $v \in V$.
- *In addition, the functions, p, μ, α, and γ are all continuously differentiable on V.*

Proof. This is Theorem 3.1 in Erlander and Smith (1990). □

This means that, if (A, C) is fully identifiable, then p, μ, α, and γ are all uniquely determined by $\mathrm{A}p = a$, and $\mathrm{C}p = c$ and (9.1).

9.2 Likelihood and Entropy of a Sample

Likelihood and entropy will be our frequent companions. We shall briefly discuss their relationship.

The likelihood of an independent sample $z = (z_1, \ldots, z_K)$ is given by

$$\mathcal{L}(z) = \Pr(z) = \prod_{k=1}^{K} p_k^{z_k}.$$

The entropy $H(p)$ of a probability distribution p is defined by

$$H(p) = -\sum_{k=1}^{K} p_k \log p_k,$$

and the observed entropy $H([z_k/N])$ of a sample of size N is given by

$$H([z_k/N]) = -\sum_{k=1}^{K} (z_k/N)\log(z_k/N).$$

From this we obtain

$$\frac{1}{N}\log\mathcal{L}(z) = \sum_{k=1}^{K}(z_k/N)\log p_k \to \sum_{k=1}^{K} p_k \log p_k = -H(p), \text{ as } N \to \infty.$$

By using the maximum likelihood estimate $\hat{p} = (z_1/N, \ldots, z_K/N)$ we obtain

$$\frac{1}{N} \log \mathcal{L}(z) \approx \sum_{k=1}^{K} (z_k/N) \log(z_k/N) = -H([z_k/N]).$$

Finally,

$$\mathcal{L}(z) \approx \exp(-NH([z_k/N])).$$

Because of this continuous monotonic functional relationship we can for large N use the observed entropy of a sample as a substitute for the likelihood of the sample.

9.3 Maximum Entropy

The entropy of a probability distribution,

$$H(p) = -\sum_{k=1}^{K} p_k \log p_k,$$

is a natural measure of dispersion or variation.

Maximizing entropy under linear constraints gives optimal solutions of the form (9.1). Hence, maximizing entropy can be used as a method for finding cost-minimizing probability distributions. In fact, we have,

Theorem 3 (Maximum Entropy). *Let the activity matrix A be rational and let the matrix B be of full rank. Then the positive probability distribution p satisfying,*

$$\sum_{k=1}^{K} p_k = 1, \quad Ap = a \quad \textit{and} \quad Cp = c, \tag{9.2}$$

is cost minimizing if and only if it is the unique optimal solution to the Maximum Entropy problem,

$$\max -\sum_{k=1}^{K} p_k \log p_k, \textit{ subject to } \sum_{k=1}^{K} p_k = 1, \quad Ap = a \quad \textit{and} \quad Cp \leq c. \tag{9.3}$$

Furthermore, the parameters μ, α and γ of the optimal solution

$$p_k = \exp(\mu + \alpha^T a_k - \gamma^T c_k), \; k = 1, \ldots, K, \tag{9.4}$$

are the Lagrange multipliers to the constraints and have the interpretation of partial derivatives, $\frac{\partial H}{\partial a_i} = \alpha_i$ and $\frac{\partial H}{\partial c_s} = \gamma_s$.

Proof. (i) Let p^m be the optimal solution to the maximum entropy problem. Then p^m is unique, because of the strict concavity of the entropy function. From the

Kuhn-Tucker conditions for the Lagrangian follows the form (9.4). It then follows from the Representation Theorem (Theorem 1) that p^m is cost minimizing. Furthermore, μ, α and γ are the Lagrange multipliers to the constraints (9.3). Also, $Cp = c$ if $\gamma > 0$.

(ii) Let p satisfying (9.2) be cost minimizing. Then, according to Theorem 1, p has the form (9.1), and hence p has the asserted form (9.4). The matrix B has full rank according to assumption. It then follows from Theorem 2, that p and the parameters μ, α and γ are unique. Both p and p^m satisfy (9.2) and have the form (9.4). Furthermore they are unique, hence they must be identical, $p = p^m$, and p is the unique optimal solution to the maximum entropy problem.

For the derivative interpretation of the Lagrange multipliers see e.g. Marlow (1978). □

9.4 Maximum Likelihood

Let $(z_1, \ldots, z_K)$ be an independent sample of size N from the cost-minimizing probability distribution

$$p_k = \exp(\mu + \alpha^T a_k - \gamma^T c_k) = \exp(\mu + \sum_{i=1}^{I} \alpha_i a_{ik} - \sum_{s=1}^{S} \gamma_s c_{sk}),\ k = 1, \ldots, K.$$

The parameters μ, α and γ can be estimated by means of the maximum likelihood method. This means finding the solutions to the problem

$$\max \sum_{k=1}^{K} z_k \log p_k \text{ subject to } \sum_{k=1}^{K} p_k = 1,$$

where

$$p_k = \exp(\mu + \alpha^T a_k - \gamma^T c_k),\ k = 1, \ldots, K.$$

Theorem 4 (Maximum Likelihood Estimates).

- *The maximum likelihood estimates $\hat{\mu}$, $\hat{\alpha}$ and $\hat{\gamma}$ of the parameters satisfy, for $i = 1, \ldots, I$, and $s = 1, \ldots, S$,*

$$\sum_{k=1}^{K} p_k = \sum_{k=1}^{K} (z_k/N) = 1, \sum_{k=1}^{K} p_k a_{ik} = \sum_{k=1}^{K} (z_k/N) a_{ik},$$

$$\sum_{k=1}^{K} p_k c_{sk} = \sum_{k=1}^{K} (z_k/N) c_{sk}. \tag{9.5}$$

- *If in addition the matrix*

$$B = \begin{Bmatrix} 1 \cdots 1 \\ A \\ C \end{Bmatrix}, \tag{9.6}$$

has full rank, then $\hat{\mu}$, $\hat{\alpha}$ *and* $\hat{\gamma}$ *are unique.*

Proof. Formula (9.5) follows from the Kuhn-Tucker conditions. Uniqueness follows from the strict concavity of the Lagrangian if the matrix B has full rank. □

If the probability distribution p has the form

$$p_k = \exp(\mu + \alpha^T a_k - \gamma^T c_k),\ k = 1, \ldots, K.$$

then the likelihood can be written

$$\prod_{k=1}^{K} p_k^{z_k} = \exp(\mu N + \sum_{i=1}^{I} \alpha_i \sum_{k=1}^{K} z_k a_{ik} - \gamma_s \sum_{k=1}^{K} z_k c_{sk}),$$

which shows that the statistic

$$\sum_{k=1}^{K} (z_k/N) a_{ik} \text{ and } \sum_{k=1}^{K} (z_k/N) c_{sk},$$

are sufficient for estimating α_i, $i = 1, \ldots, I$, and γ_s, $s = 1, \ldots, S$ (see e.g. Lehmann 1959, p. 51).

The normalizing parameter μ can be excluded from the formulation, since it is just a constant that makes the probabilities sum up to one,

$$p_k = \frac{\exp(\alpha^T a_k - \gamma^T c_k)}{\sum_{k=1}^{K} \exp(\alpha^T a_k - \gamma^T c_k)}.$$

Let $\beta = [\alpha, \gamma]$. Under regularity conditions, as $N \to \infty$, $\sqrt{N}(\hat{\beta} - \beta)$ is asymptotically normal with mean zero and covariance matrix $[I(\beta)]^{-1}$, where $I(\beta)$ is the information matrix

$$I_{st}(\beta) = -E[\frac{\partial^2 \log p_k}{\partial \beta_s \partial \beta_t}].$$

(See e.g. Lehmann 1983, Theorem 6.4.1 p. 429, Ex. 6.5.3 p. 438.)

References

Abrahamsson T, Lundqvist L (1999) Formulation and estimation of combined network equilibrium models with applications to Stockholm. Transp Sci 33:80–100

Anderson SP, de Palma A, Thisse JF (1992) Discrete choice theory of product differentiation. MIT, Cambridge, Massachusetts

Arrow KJ (1995) A note on freedom and flexibility. In: Basu K, Pattanaik P, Sysumura K (eds) Choice, welfare and development. Clarendon, London

Bachem A, von Randow R (1978) Integer theorems of farkas lemma type. Oper Res Verfahr 32: 19–28

de la Barra T (1998) Improved logit formulations for integrated land use, transport and environmental models. In: Lundqvist L, Mattsson LG, Kim TJ (eds) Network infrastructure and the urban environment. Springer, Berlin, Heidelberg, New York, pp 288–307

Beckmann MJ, McGuire CB, Winsten CB (1956) Studies in the economics of transportation. Yale University Press, New Haven

Bell MGH, Iida Y (1997) Transportation network analysis. Wiley, Chichester

Ben-Akiva M (1974) Structure of passenger travel demand models. Transp Res Rec 526:26–42

Ben-Akiva M, Lerman SR (1985) Discrete choice analysis: theory and application to travel demand. MIT, Cambridge, Massachusetts

Bliemer MCJ, Rose JM, Hensher DA (2009) Efficient stated choice experiments for testing nested logit models. Transp Res 43B:19–35

Boersch-Supan A, Pitkin J (1982) Multinomial logit models of housing choices. Working Paper No 79, Center fo Urban Studies. MIT and Harvard, Cambridge Massachusetts

Boyce D (1998) Long-term advances in the state of the art of travel forecasting methods. In: Marcotte P, Nguyen S (eds) Equilibrium and advanced transportation modelling. Kluwer, Dordrecht, pp 73–86

Boyce DE (1980) A framework for constructing network equilibrium models of urban location. Transp Sci 14:77–96

Boyce DE (1984) Urban transportation network equilibrium and design models: recent achievements and future prospects. Environ Plan 16A:1445–1474

Boyce DE (2007) Forecasting travel on congested urban transportation networks: review and prospects for network equilibrium models. Netw Spat Econ 7:99–128

Boyce DE, Williams HW (2005) Urban travel forecasting in the USA and UK. In: Reggiani A, Schintler L (eds) Methods and models in transport and telecommunications. Springer, Berlin, pp 25–44

Boyce DE, LeBlanc LJ, Chon KS, Lee YJ, Lin KT (1983) Implementation and computational issues for combined models of location, destination, mode and route choice. Environ Plan A 15:1219–1230

Boyce DE, LeBlanc LJ, Chon KS (1988) Network equilibrium models of urban location and travel choices: a retrospective survey. J Reg Sci 28:159–183

Bregman LM (1967) Proof of the convergence of Sheleikhovskii's method for a problem with transportation constratints. USSR Comput Math Math Phys 2:191–204
Brundell-Freij K (1995) The logit model applied to the modal split of regional computing. Technical report, PhD thesis, Department of Traffic Planning and Engineering, University of Lund, Lund, Sweden
Carey HC (1858) Principles of social science. Lippincott, Philadelphia
Daganzo CF, Sheffi Y (1977) On stochastic models of traffic assignment. Transp Sci 11:253–274
Daly A (1982) Applicability of disaggregate models of behaviour: a question of methodology. Transp Res 16A:363–370
Daly AJ, Zachary S (1978) Improved multiple choice models. In: Hensher DA, Dalvi MQ (eds) Determinants of travel choice. Saxon House, Sussex, pp 335–357
Damberg O, Lundgren JT, Patriksson M (1996) An algorithm for the stochastic user equilibrium problem. Transp Res B 30:115–131
Dantzig GB (1963) Linear programming and extensions. Princeton University Press, Princeton, NJ
Erlander S (1975) Applications of mathematical and statistical models in the field of road traffic and safety. Technical Report LiH-MAT-R-75-1, Department of Mathematics, Linköping Institute of Technology
Erlander S (1977) Accessibility, entropy and distribution and assignment of traffic. Transp Res 11:149–153
Erlander S (1980) Optimal spatial interaction and the gravity model, lecture notes in economics and mathematical systems, vol 173. Springer, Berlin, Heidelberg, New York
Erlander S (1985) On the principle of monotone likelihood and loglinear models. Math Programming 21:137–151
Erlander S (1990) Efficient population behavior and the simultaneous choices of origins, destinations and routes. Transp Res 24B:363–373
Erlander S (1998) Efficiency and the logit model. Ann Oper Res 82:203–218
Erlander S (2002) Linear-in-parameters logit model derived from the efficiency principle. In: Gendreau M, Marcotte P (eds) Transportation and network analysis: current trends. Kluwer, Dordrecht, pp 109–117
Erlander S (2005) Welfare, freedom of choice and composite utility in the logit model. Soc Choice Welfare 24:509–525
Erlander S, Lundgren JT (2004) Cost minimizing behavior in random discrete choice modeling. In: Lee DH (ed) Urban and regional transportation modeling: essays in honor of David Boyce, new dimensions in networks. Edward Elgar, Northampton, Massachusetts, USA, pp 70–82
Erlander S, Schéele S (1974) A mathematical programming model for bus traffic in a network. In: Buckley DJ (ed) Transportation and traffic theory. Elsevier, New York, pp 581–605
Erlander S, Smith TE (1990) General representation theorems for efficient population behavior. Appl Math Comput 36:173–217
Erlander S, Stewart NF (1990) The gravity model in transportation analysis – theory and extensions. VSP, Utrecht, The Netherlands
Evans AW (1970) Some properties of trip distribution methods. Transp Res 4:19–36
Evans SP (1973) A relationship between the gravity model for trip distribution and the transportation problem in linear programming. Transp Res 7:39–61
Evans SP (1976) Derivation and analysis of some models for combining trip distribution and assignment. Transp Res 10:37–57
Fisk C (1980) Some developments in equilibrium traffic assignment. Transp Res 14B:243–255
Fisk CS, Boyce DE (1984) A modified composite cost measure for probabilistic choice modeling. Environ Plan A 16:241–248
Florian M, Nguyen S, Ferland J (1975) On the combined distribution-assignment of traffic. Transp Sci 9:43–53
Florian M, Chapleau R, Nguyen S (1977) Emme: a planning method for multi-modal urban transportation systems, Publication 62. Centre de Recherche sur les Transport, Montreal, Canada
Fry TRL, Harris MN (1998) Testing for independence of irrelevant alternatives. Sociol Methods Res 26:401–423

Gärling T (1998) Theoretical foundations of travel choice modeling: An introduction. In: Gärling T, Laitila T, Westin K (eds) Theoretical foundations of travel choice modeling. Elsevier, pp 1–30
Greene WH, Hensher DA (2003) A latent class model for discrete choice analysis. Transp Res B 37:681–698
Gumbel EJ (1958) Statistics of Extremes. Columbia University Press, New York
Hausman JA, McFadden D (1984) Specification tests for the multinomial logit model. Econometrica 52:1219–1240
Hensher DA, Button KJ (eds) (2000) Handbook of transport modelling. Pergamon, Amsterdam
Hensher DA, Greene WH (2002) Specification and estimation of the nested logit model: alternative normalisations. Transp Res 36B:1–17
Hjorth U (1994) Computer intensive statistical methods. Chapman and Hall, London
Jara-Diaz SR (2000) Allocation and valuation of travel-time savings. In: Hensher DA, Button KJ (eds) Handbook of transport modelling. Pergamon, Amsterdam, pp 303–319
Kantorovich LV (1939) Mathematicheskie metody organisatsii i planirovania porizvodsta. Leningrad State University Publishers, St. Petersburg
Koopmans TC (1951) Optimum utilization of the transportation system. In: Proceedings of the international statistical conference, vol 5, pp 135–145
Koppelman FS, Wen CH (1998) Alternative nested logit models: structure, properties and estimation. Transp Res B 32:289–298
Kruithof J (1937) Telefoonverkehrsrekening. De Ingenieur 52:E15–E25, calculation of telephone traffic. Translation No 2663, Post Office Research Library, London
Kullback S (1959) Information theory and statistics. Wiley, New York
Kulldorff G (1955) Migration probabilities. Technical Report Series B: No 14. Department of Geography, Lund University, Lund, Sweden
Lehmann EL (1959) Testing statistical hypotheses. Wiley, New York
Lehmann EL (1983) Theory of point estimation. Wiley, New York
Lill E (1891) Das reisegesetz und seine anwendung auf den eisenbahnverkehr. Wien
Luce RD (1959) Individual choice behavior: a theoretical analysis. Wiley, New York
Luce RD, Suppes P (1965) Preference, utility and subjective probability. In: Luce RD, Bush RR, Galanter E (eds) Handbook of mathematical psychology, vol 3. Wiley, pp 249–410
Lundgren JT (1989) Optimization approaches to travel demand modelling. PhD thesis, Department of Mathematics, Linköping Institute of Technology, Linköping, Sweden
Lundgren JT, Patriksson M (1998) The combined distribution and stochastic assignment problem. Ann Oper Res 82:309–329
Marlow WH (1978) Mathematics for operations research. Wiley, New York
Mattsson LG (1984) Equivalence between welfare and entropy approaches to residential location. Reg Sci Urban Econ 14:147–173
Mattsson LG, Weibull JW (2002) Probabilistic choice and procedurally bounded rationality. Games Econ Behav 41:61–78
McFadden D (1974) Conditional logit analysis of qualitative choice behavior. In: Zarembka P (ed) Frontiers in econometrics. Academic, New York, pp 105–142
McFadden D (1978a) Modeling the choice of residential location. In: Karlqvist A, Lundqvist L, Snickars F, Weibull JW (eds) Spatial interaction theory and planning models. North-Holland, Amsterdam, pp 75–96
McFadden D (1978b) Quantitative methods for analyzing travel behavior of individuals, some recent developments. In: Hensher D, Stopher P (eds) Behavioral travel modelling. Croom Helm, London, pp 279–318
McFadden D (1981) Econometric models of probabilistic choice. In: Manski C, McFadden D (eds) Structural analysis of discrete data with econometric applications. MIT, Cambridge, Massachusetts, pp 198–272
McFadden D (1987) Regression-based specification tests for the multinomial logit model. J Econom 34:63–82
McFadden D (1999) Rationality for economists? J Risk Uncertain 19:73–105

McFadden D (2001) Disaggregate behavioral travel demand's RUM side – a 30-year retrospective. In: Hensher D (ed) Travel behavior research. The Leading Edge, Amsterdam, Pergamon, pp 17–64
McFadden D, Train K (2000) Mixed MNL models for discrete response. J Appl Econ 15:447–470
McFadden D, Tye W, Train K (1977) An application of diagnostic tests for the irrelevant alternatives properties of the multinomial logit model. Transp Res Rec 637:39–46
McFadden D, Train K, Tye WB (1981) An application of diagnostic tests for the independence of irrelevant alternatives of the multinomial logit model. Transp Res Rec 637:39–46
Miyagi T, Morisugi H (1996) A direct measure of the value of choice-freedom. Pap Reg Sci J RSAI 75:121–134
Monge G (1781) Mémoire sur la théorie des déblai et des remblais. Mémoires de l académie royale des sciences Paris pp 666–704
Murchland JD (1966) Some remarks on the gravity model of traffic distribution and an equivalent maximization formulation. Technical Report LSE-TNT-38, Transport network theory unit, London School of Economics, London
Ortuzar JD (2001) On the development of the nested logit model. Transp Res B 35:213–216
Patriksson M (1994) The traffic assignment problem. VSP, Utrecht, Tokyo
Popper KR (1963) Conjectures and refutations: the growth of scientific knowledge. Routledge, London
Rabin M (1998) Psychology and economics. J Econ Lit 36:11–46
Ravenstein EG (1985) The laws of migration. J Stat Soc 48
Rosenthal RW (1973) The network equilibrium problem in integers. Networks 3:53–59
Roy JR, Thill JC (2004) Spatial interaction modelling. Pap Reg Sci 83:339–361
Schéele S (1977) A mathematical programming algorithm for optimal bus frequencies. PhD thesis, Linköping University, Linköping, Sweden
Schéele S (1980) A supply model for public transit services. Transp Res B pp 133–146
Sen A (1985) Commodities and capabilities. Oxford University Press, New Delhi
Sen A (1988) Freedom of choice: concept and content. Eur Econ Rev 32:269–294, 960418
Sen A (1991) Welfare, preference and freedom. J Econom 50:15–29
Sen A (1992) Inequality reexamined. Clarendon, Oxford
Sen A, Smith TE (1995) Gravity models of spatial interaction behavior. Springer
Sen AK (1995) Maximizing behaviour and the choice act. World Econometric Congress in Tokyo
Shannon CE (1948) A mathematical theory of communication. Bell Syst Tech J 27:379–423, 623–656
Sheffi Y (1985) Urban transportation networks, equilibrium analysis with mathematical programming methods. Prentice-Hall, Englewood Cliffs
Small KA (1994) Approximate generalized extreme value models of discrete choice. J Econom 62:351–382
Small KA, Hsiao C (1985) Multinomial logit specification tests. Int Econ Rev 26:619–627
Smith TE (1978a) A cost-efficiency principle of spatial interaction behavior. Reg Sci Urban Econ 8:313–337
Smith TE (1978b) A general efficiency principle of spatial interaction. In: Karlqvist A, Lundqvist L, Snickars F, Weibull JW (eds) Spatial interaction theory and planning models. North Holland, Amsterdam, pp 97–118
Smith TE (1983) A cost-efficiency approach to the analysis of congested spatial-interaction behavior. Environ Plan A 15:435–464
Smith TE (1985) Remarks on the most-probable-state approach to analyzing probabilistic theories of behavior. Environ Plan A 17:688–695
Smith TE (1988) A cost-efficiency theory of dispersed network equilibria. Environ Plan A 20: 231–266
Smith TE (1990) Most-probable-state analysis: a method for testing probabilistic theories of population behavior. In: Chatterji M, Kuenne RE (eds) New frontiers in regional science. MacMillan, London, pp 75–94

Suppes P (1987) Maximizing freedom of decision: an axiomatic approach. In: Arrow and the foundations of the theory of economic policy. New York University Press, New York, pp 243–254

Suppes P (1996) The nature and measurement of freedom. Soc Choice Welfare 13:183–200

Tomlin JA (1967) Mathematical programming models for traffic network problems. Doctorial dissertation, University of Adelaide, Australia

Varian H (1992) Microeconomic analysis. Norton, New York

Wardrop JG (1952) Some theoretical aspects of road traffic research. In: Proceedings of the Institution of Civil Engineers, Part II, vol 1, pp 325–378

Weibull W (1939) A statistical theory of the strength of materials. Technical Report, No 151. Royal Swedish Academy of Engineering Sciences, Stockholm

Williams HCWL (1977) On the formation of travel demand models and economic evaluation measures of user benefit. Environ Plan A 9:285–344

Williams HCWL, Senior ML (1977) Model-based transport policy assessment: 2. removing fundamental inconsistencies from the models. Traffic Eng Control 18:464–469

Wilson AG (1967) A statistical theory of spatial distribution models. Transp Res 1:253–269

Wilson AG (1970) Entropy in urban and regional modelling. Pion, London

Wilson AG, Coelho JD, MacGill SM, Williams HCWL (1981) Optimization in locational and transport analysis. Wiley, Chichester

FSC
www.fsc.org
MIX
Papier | Fördert
gute Waldnutzung
FSC® C083411

Kolibri 360 GmbH
Ferdinand-Jühlke-Straße 7
99095 Erfurt, Deutschland
produktsicherheit@kolibri360.de